Topological Groups

Topological Groups

Advances, Surveys, and Open Questions

Special Issue Editor

Sidney A. Morris

MDPI • Basel • Beijing • Wuhan • Barcelona • Belgrade

MDPI

Special Issue Editor
Sidney A. Morris
La Trobe University and
Federation University Australia
Australia

Editorial Office
MDPI
St. Alban-Anlage 66
4052 Basel, Switzerland

This is a reprint of articles from the Special Issue published online in the open access journal *Axioms* (ISSN 2075-1680) from 2017 to 2019 (available at: https://www.mdpi.com/journal/axioms/special_issues/topological_groups).

For citation purposes, cite each article independently as indicated on the article page online and as indicated below:

LastName, A.A.; LastName, B.B.; LastName, C.C. Article Title. *Journal Name* **Year**, *Article Number*, Page Range.

ISBN 978-3-03897-644-8 (Pbk)
ISBN 978-3-03897-645-5 (PDF)

Contents

About the Special Issue Editor

Sidney Morris, Adjunct Professor, La Trobe University and Emeritus Professor, Federation University, Australia. He received BSc (Hons) from University of Queensland in 1969 and Ph.D. from Flinders University in 1970. He held positions of Professor, Department Head, Dean, Vice-President, CAO, and CEO. He has been employed by 13 universities in Australia, UK, USA, and Israel. He was Editor of the Bulletin of Australian Mathematical Society and Journal of Research and Practice in Information Technology, and founding editor of Australian Mathematical Society Lecture Series and Journal of Group Theory. He served on the Council of the Australian Mathematical Society for 25 years and as Vice-President, and received the Lester R. Ford Award from Mathematical Association of America. He has published 160 refereed journal papers and four books for undergraduates, graduates, and researchers, plus an online book, translated into eight languages. In 2016 he edited the book "Topological Groups: Yesterday, Today, Tomorrow".

Preface to "Topological Groups"

In 1900 David Hilbert presented an address at the International Congress of Mathematicians in Paris and formulated 23 problems that influenced much of the research in the 20th century. The fifth of these problems asked whether every locally euclidean group admits a Lie group structure. This motivated a great amount of research on locally compact groups, culminating in the 1950s with the work of Gleason, Iwasawa, Montgomery, Yamabe, and Zippin, which gave a positive answer to Hilbert's question and developed much structure theory of locally compact groups to boot. In the 1940s, the work on the free topological groups of Markov and Graev expanded the study of topological groups in a serious way to non-locally compact groups. In the early 21st century, pro-Lie groups were introduced as a natural and well-behaved extension of the notions of connected locally compact groups and compact groups focusing on their Lie theory. The Special Issue of Axioms called "Topological Groups: Yesterday, Today, Tomorrow" was published as book in 2016 and has had a tremendous reception. It addressed some of the significant research of this 115-year period. A wonderful feature of the book was the inclusion of surveys and a large number of open questions.

This second volume on topological groups, called "Topological Groups: Advances, Surveys, and Open Questions", contains recent articles by some of the best scholars in the world on topological groups including Taras Banakh, Michael Megrelishvili, George A. Willis, Dmitri Shakhmatov, and O'lga V. Sipacheva, as well as a paper by the renowned scholar Saharon Shelah. A feature of the first volume was the surveys, and we continue that tradition in this second volume with three new surveys. These surveys are of interest, not only to the expert but also to those who are less experienced. Particularly exciting to active researchers, especially young researchers, is the inclusion of over three dozen open questions. This volume consists of 11 papers containing many new and interesting results and examples across the spectrum of topological group theory and related topics.

The first paper in this book is "Separability of Topological Groups: A Survey with Open Problems" by Arkady Leiderman and Sidney A. Morris. In recent years, Leiderman has been a leader in the study of the separability of topological groups. This paper alone states 20 open questions and puts them in context.

The second paper, "Categorically Closed Topological Groups", is also a survey, this time by Taras Banakh. This paper surveys existing and new results on topological groups that are C-closed for various categories C of topologized semigroups. In particular, it analyzes solutions to a general problem consisting of 45 subproblems.

The third paper is "Selective Survey on Spaces of Closed Subgroups of Topological Groups" by Igor V. Protasov. This paper surveys the Chabauty topology and the Vietoris topology on the set of all closed subgroups of a topological group, and in the author's words "...is my subjective look at this area".

The fourth paper, "No Uncountable Polish Group Can be a Right-Angled Artin Group", is by Gianluca Paolini and Saharon Shelah. Generalizing results on free groups and free abelian groups, the authors prove that the automorphism group of a countable structure cannot be an uncountable right-angled Artin group.

The fifth paper is "Computing the Scale of an Endomorphism of a Totally Disconnected Locally Compact Group" by George A. Willis. While a substantial amount of information about the structure of connected locally compact groups, even almost connected locally compact groups, has been known for over half a century, little of substance was known about totally disconnected locally compact groups before the deep contributions of Willis. In this paper, the scale of an endomorphism of a totally disconnected locally compact group is defined and "the information required to compute the scale is reviewed from the perspective of the, as yet incomplete, general theory of totally disconnected, locally compact groups".

The sixth paper is "Extending Characters of Fixed Point Algebras" by Stefan Wagner. Given a dynamical system (A, G, a) with a complete commutative continuous inverse algebra A and a compact group G, it is shown that each character of the corresponding fixed point algebra can be extended to a character of A.

The seventh paper is "A Note on the Topological Group c_0" by Michael Megrelishvili. It is shown that Gromov's compactification of c_0 is not a semigroup compactification. The paper also contextualizes three open questions.

The eighth paper is "Large Sets in Boolean and Non-Boolean Groups and Topology" by Ol'ga V. Sipacheva. "Various notions of large sets in groups, including the classical notions of thick, syndetic, and piecewise syndetic sets, and the new notion of vast sets in groups, are studied, with an emphasis on the interplay between such sets in Boolean groups. Natural topologies closely related to vast sets are considered; as a byproduct, interesting relations between vast sets and ultrafilters are revealed."

The ninth paper is "Selectively Pseudocompact Groups without Infinite Separable Pseudocompact Subsets" by Dmitri Shakhmatov and Víctor Hugo Yañez. This paper answers an open question by producing a ZFC example of a selectively pseudocompact (abelian) group that is not selectively sequentially pseudocompact. This leaves open the question: is there a ZFC example of a countably compact (abelian) group that is not selectively sequentially pseudocompact? The authors also show that that the free precompact Boolean group of a topological sum of spaces, each of which is either maximal or discrete, contains no infinite separable pseudocompact subsets. The authors also state another open question.

The tenth paper is "(L)-Semigroup Sums" by John R. Martin. Noting that an (L)-semigroup S is a compact n-manifold with connected boundary B together with a monoid structure on S such that B is a subsemigroup of S, the author shows that no (L)-semigroup sum of dimension less than or equal to five admits an H-space structure, and that such sums cannot be a retract of a topological group.

The eleventh and final paper is "Varieties of Coarse Spaces" by Igor Protasov. The study of varieties of groups has its roots in the 1930s with the work of Garrett Birkoff and B.H. Neumann. The study of varieties of topological groups began with a series of papers by Sidney A. Morris, the first of which appeared in 1969 and resulted from research he began as an undergraduate under the supervision of Ian D. Macdonald. This paper is a natural extension of that work. A class of coarse spaces is called a variety if it is closed under the formation of subspaces, coarse images, and products. The author classifies the varieties of coarse spaces and, in particular, shows that if a variety contains an unbounded metric space then it is the variety of all coarse spaces.

Sidney A. Morris
Special Issue Editor

axioms

MDPI

Article

Separability of Topological Groups: A Survey with Open Problems

Arkady G. Leiderman [1,*,†] and **Sidney A. Morris** [2,3,†]

[1] Department of Mathematics, Ben-Gurion University of the Negev, P.O. Box 653, Beer Sheva 84105, Israel
[2] Centre for Informatics and Applied Optimization (CIAO), Federation University Australia, P.O. Box 663, Ballarat 3353, Australia; morris.sidney@gmail.com
[3] Department of Mathematics and Statistics, La Trobe University, Melbourne 3086, Australia
[*] Correspondence: arkady@math.bgu.ac.il; Tel.: +972-8-647-7811
[†] These authors contributed equally to this work.

Received: 25 November 2018; Accepted: 25 December 2018; Published: 29 December 2018

Abstract: Separability is one of the basic topological properties. Most classical topological groups and Banach spaces are separable; as examples we mention compact metric groups, matrix groups, connected (finite-dimensional) Lie groups; and the Banach spaces $C(K)$ for metrizable compact spaces K; and ℓ_p, for $p \geq 1$. This survey focuses on the wealth of results that have appeared in recent years about separable topological groups. In this paper, the property of separability of topological groups is examined in the context of taking subgroups, finite or infinite products, and quotient homomorphisms. The open problem of Banach and Mazur, known as the Separable Quotient Problem for Banach spaces, asks whether every Banach space has a quotient space which is a separable Banach space. This paper records substantial results on the analogous problem for topological groups. Twenty open problems are included in the survey.

Keywords: separable topological group; subgroup; product; isomorphic embedding; quotient group; free topological group

---

1. Introduction

All topological spaces and topological groups are assumed to be Hausdorff and all topological spaces are assumed to be infinite unless explicitly stated otherwise.

The fundamental topological operations which produce new topological groups from given ones are:

(1) taking subgroups;
(2) taking finite or infinite products;
(3) open continuous homomorphic images = quotient images;
(4) (topological group) isomorphic embeddings.

A topological space which has a dense countable subspace is called *separable*.

The main aim of this survey paper is to present systematically the results concerning the behavior of separability of topological groups with respect to the topological operations listed above and make clear which problems are open. Much of the material is from the recent publications [1–8].

Informally speaking, this survey contributes to the manifestation of the phenomenon that the structure of topological groups is much more sensitive to the presence of countable topological properties than is the structure of general topological spaces.

Section 2 sketches the relevant background results about separability of general topological spaces. Section 3 is devoted to the closed subgroups of separable topological groups and isomorphic embeddings into separable topological groups. Section 4 is devoted to the products of separable topological groups and the products of separable topological vector spaces. Section 5 is devoted to the separable quotient problem for general topological groups. Section 6 is devoted to the metrizable and separable quotients of free topological groups. Section 7 deals with the question of when an abstract group can be equipped with a separable topological group topology.

At the end of most sections we pose open problems, 20 problems in all. Throughout the paper anything labeled as a Problem is an open problem. We also have many Questions, for which an answer is provided.

The reader is advised to consult the monographs of Engelking [9] and Arhangel'skii and Tkachenko [10] for any notions which are not explicitly defined in our paper.

2. Separability of Topological Spaces

The *weight* $w(X)$ of a topological space X is defined as the smallest cardinal number $|\mathcal{B}|$, where $\mathcal{B}$ is a base of the topology on X. The *density character* $d(X)$ of a topological space X is $\min\{|A| : A \text{ is dense in } X\}$. Recall that if $d(X) \leq \aleph_0$, then we say that the space X is *separable*. We denote by $\mathfrak{c}$ the cardinality of continuum.

A topological space X is said to be *hereditarily separable* if X and every subspace of X is separable. A topological space is said to be *second countable* if its topology has a countable base. A topological space X is said to have a *countable network* if there exists a countable family $\mathcal{B}$ of (not necessarily open) subsets such that each open set of X is a union of members of $\mathcal{B}$.

- Any space with a countable network is hereditarily separable;
- a metrizable space is separable if and only if it is second countable;
- any continuous image of a separable space is separable;
- countable networks are preserved by continuous images.

2.1. Weight of Separable Topological Spaces

Theorem 1. (De Groot, ([11], Theorem 3.3)) *If X is a separable regular space, then $w(X) \leq \mathfrak{c}$. More generally, every regular space X satisfies $w(X) \leq 2^{d(X)}$, and then $|X| \leq 2^{w(X)} \leq 2^{2^{d(X)}}$.*

Compact dyadic spaces are defined to be continuous images of generalized Cantor cubes $\{0,1\}^\kappa$, where κ is an arbitrary cardinal number. It is well-known ([12], Theorem 10.40) that every compact group is dyadic.

Proposition 1. (Engelking, ([13], Theorem 10)) *Let κ be an infinite cardinal. A compact dyadic space K with $w(K) \leq 2^\kappa$ satisfies $d(K) \leq \kappa$. In particular, if $w(K) \leq \mathfrak{c}$, then K is separable.*

2.2. Products of (Hereditarily) Separable Topological Spaces

Theorem 2. (Hewitt–Marczewski–Pondiczery, [9]) *Let $\{X_i : i \in I\}$ be a family of topological spaces and $X = \prod_{i \in I} X_i$, where $|I| \leq 2^\kappa$ for some cardinal number $\kappa \geq \omega$. If $d(X_i) \leq \kappa$ for each $i \in I$, then $d(X) \leq \kappa$. In particular, the product of no more than $\mathfrak{c}$ separable spaces is separable.*

Remark 1. The Sorgenfrey line $\mathbb{S}$ is a hereditarily separable space whose square $\mathbb{S} \times \mathbb{S}$ has the uncountable discrete subspace $\{(x, -x) : x \in \mathbb{S}\}$ as a subspace and so $\mathbb{S} \times \mathbb{S}$ is not hereditarily separable.

Proposition 2 ([1]). *Let X be a hereditarily separable space and Y a space with a countable network. Then the product $X \times Y$ is also hereditarily separable.*

2.3. Closed Embeddings into Separable Topological Spaces

Any compact space of weight not greater than $\mathfrak{c}$ homeomorphically embeds into the separable compact cube $[0, 1]^{\mathfrak{c}}$.

A Tychonoff space X is called *pseudocompact* if every continuous real-valued function defined on X is bounded. Similarly to Theorem 11 one can prove:

Proposition 3. *Every Tychonoff space of weight not greater than $\mathfrak{c}$ is homeomorphic to the closed subspace of a separable pseudocompact space.*

2.4. Separable Quotient Spaces of Topological Spaces

Assume $\varphi \colon X \to Y$ is a mapping such that (1) φ is surjective, (2) φ is continuous, and (3) for $U \subseteq Y$, $\varphi^{-1}(U)$ is open in X implies that U is open in Y. In this case, the mapping φ is called a *quotient* mapping.

Every closed mapping and every open mapping is a quotient mapping.

The majority of topological properties are not preserved by quotient mappings. For instance, a quotient space of a metric space need not be a Hausdorff space, and a quotient space of a separable metric space need not have a countable base.

A surjective continuous mapping $\varphi \colon X \to Y$ is said to be *R-quotient* [14] if for every real-valued function f on Y, the composition $f \circ \varphi$ is continuous if and only if f is continuous. Clearly, every quotient mapping is R-quotient, but the converse is false.

Let $\varphi \colon X \to Y$ be a surjective continuous mapping, where the space Y is Tychonoff. Then Y admits the finest topology, say, σ such that the mapping $\varphi \colon X \to (Y, \sigma)$ is R-quotient. The topology σ of Y is initial with respect to the family of real-valued functions f on Y such that the composition $f \circ \varphi$ is continuous. It is easy to see that the space (Y, σ) is also Tychonoff and that σ is finer than the original topology of Y. We say that σ is the *R-quotient topology* on Y (with respect to φ). Notice that the mapping $\varphi \colon X \to (Y, \sigma)$ remains continuous.

Proposition 4 ([2]). *Every continuous mapping of a pseudocompact space onto a first countable Tychonoff space is R-quotient. Therefore for every pseudocompact space there exists an R-quotient mapping onto infinite subset of the closed unit interval $[0, 1]$. Every locally compact space also admits an R-quotient mapping onto an infinite subset of the closed unit interval $[0, 1]$.*

Recall that the class of Lindelöf Σ-spaces is the smallest class of topological spaces which contains all compact and all separable metrizable topological spaces, and is closed with respect to countable products, closed subspaces, and continuous images (see [10], Section 5.3).

Proposition 5 ([2]). *For every Lindelöf Σ-space (in particular, σ-compact space) X there exists an R-quotient mapping onto an infinite space with a countable network.*

2.5. Open Problems

Problem 1 ([2]). *Does there exist an R-quotient mapping from each Tychonoff space onto an infinite subspace of $[0, 1]$?*

Furthermore, a more particular question below is open:

Problem 2 ([2]). *Does there exist an R-quotient mapping from each Lindelöf space onto an infinite separable Tychonoff space?*

3. Subgroups of Separable Topological Groups

3.1. Topological Groups with a Dense Compactly Generated Subgroup

A topological group G is said to be *compactly generated* if it has a compact subspace K such that the smallest subgroup of G which contains K is G itself. The topological group G is said to be *finitely generated modulo open sets* if for every open set $U \subseteq G$, there exists a finite set $F \subset G$ such that the smallest subgroup of G which contains $F \cup U$ is G itself.

Since every metrizable compact space is separable, it is easy to see that every metrizable topological group which has a dense compactly generated subgroup is separable. The next theorem says under what conditions the converse is also true.

Theorem 3 ([15]). *A metrizable topological group G has a dense compactly generated subgroup if and only if it is separable and finitely generated modulo open sets.*

Corollary 1. *Let G be a metrizable connected topological group. Then G is separable if and only if it has a dense compactly generated subgroup.*

Corollary 2. *Let G be an additive topological group of a metrizable topological vector space. Then G is separable if and only if it has a dense compactly generated subgroup.*

3.2. Characterization of Subgroups of Separable Topological Groups

A topological group is said to be *ω-narrow* ([10], Section 3.4) if it can be covered by countably many translations of every neighborhood of the identity element. It is known that every separable topological group is ω-narrow (see [10], Corollary 3.4.8). The class of ω-narrow groups is productive and hereditary with respect to taking arbitrary subgroups ([10], Section 3.4), so ω-narrow groups need not be separable. In fact, an ω-narrow group G can have uncountable cellularity, i.e., there is an uncountable family of disjoint non-empty open subsets in G (see [10], Example 5.4.13).

The following theorem characterizes the class of ω-narrow topological groups.

Theorem 4. (Guran, [16]) *A topological group G is ω-narrow if and only if it is topologically isomorphic to a subgroup of a product of second countable topological groups.*

A topological group which has a local base at the identity element consisting of open subgroups is called *protodiscrete*. A complete protodiscrete group is said to be *prodiscrete*. Protodiscrete topological groups are exactly the totally disconnected pro-Lie groups ([17], Proposition 3.30).

Theorem 5 ([4]). *A (protodiscrete abelian) topological group H is topologically isomorphic to a subgroup of a separable (prodiscrete abelian) topological group if and only if H is ω-narrow and satisfies $w(H) \leq \mathfrak{c}$.*

It is natural to compare the restrictions on a given topological group G imposed by the existence of either a topological embedding of G into a separable regular space or a topological isomorphism of G onto a subgroup of a separable topological group.

Let us note that the first of the two classes of topological groups is strictly wider than the second one. In order to show this, consider an arbitrary discrete group G satisfying $\omega < |G| \leq \mathfrak{c}$. Then G embeds as a *closed subspace* into the separable space $\mathbb{N}^{\mathfrak{c}}$ [9], where $\mathbb{N}$ is the set of positive integers endowed with the discrete topology. However, G does not admit a topological isomorphism onto a subgroup of a separable topological group. Indeed, every subgroup of a separable topological group is ω-narrow by Theorem 5. Since the discrete group G is uncountable, it fails to be ω-narrow.

The above observation makes it natural to restrict our attention to ω-narrow topological groups when considering embeddings into separable topological groups. It turns out that in the class of

ω-narrow topological groups, the difference between the two types of embeddings disappears, even if we require an embedding to be closed.

In the next result, which complements Theorem 5, we identify a large class of topological groups with the class of closed subgroups of separable path-connected, locally path-connected topological groups.

Theorem 6 ([4]). *The following are equivalent for an arbitrary ω-narrow topological group G:*

(a)   *G is homeomorphic to a subspace of a separable regular space;*
(b)   *G is topologically isomorphic to a subgroup of a separable topological group;*
(c)   *G is topologically isomorphic to a closed subgroup of a separable path-connected, locally path-connected topological group.*

Next, we consider the following question: Let G be a separable topological group. Under what conditions is every closed subgroup of G separable?

Historically, the first non-trivial result is due to Itzkowitz [18].

Theorem 7. *Let G be a separable compact topological group. Then every closed subgroup of G is separable.*

Please note that σ-compactness of G is not a sufficient condition.

Example 1. Let X be any separable compact space which contains a closed non-separable subspace Y. The free abelian topological group $A(Y)$ naturally embeds into $A(X)$ as a closed subgroup. Then $A(X)$ is a separable σ-compact group, while $A(Y)$ is not separable—otherwise Y would be separable (see [19], Lemma 3.1).

Let us recall that a topological group G is called *feathered* if it contains a non-empty compact subset with a countable neighborhood base in G. Equivalently, G is feathered if it contains a compact subgroup K such that the quotient space G/K is metrizable (see [10], Section 4.3). All metrizable groups and all locally compact groups are feathered. Notice also that the class of feathered groups is closed under taking countable products (see [10], Proposition 4.3.13).

Theorem 8 ([4]). *Let a feathered topological group G be isomorphic to a subgroup of a separable topological group. Then G is separable.*

Since the class of feathered topological groups includes both locally compact and metrizable groups, Theorem 8 provides a generalization of the results of Comfort and Itzkowitz [19] for locally compact groups and also well-known results for metrizable groups [20,21].

Corollary 3. *If a locally compact topological group G is isomorphic to a subgroup of a separable topological group, then G is separable.*

Corollary 4. *If a metrizable group G is isomorphic to a subgroup of a separable topological group, then G is separable.*

Recall that a non-empty class Ω of topological groups is said to be a *variety* [22–27] if it is closed under the operations of taking subgroups, quotient groups, (arbitrary) cartesian products and isomorphic images. Let $\mathcal{C}$ be a class of topological groups and let $\mathfrak{V}(\mathcal{C})$ be the intersection of all varieties containing $\mathcal{C}$. Then $\mathfrak{V}(\mathcal{C})$ is said to be the *variety generated by* $\mathcal{C}$. With the help of results from [26] Corollary 4 can be extended as follows.

Corollary 5. *If C is any class of separable abelian topological groups, then every metrizable group in $\mathfrak{V}(C)$ is separable.*

It is clear that Corollary 5 would be false if the metrizability condition is deleted or replaced by feathered or even compact.

Problem 3. *Does Corollary 5 remain valid if we drop the assumption that all groups in the class C are abelian?*

Remark 2 ([4]). (1) A discrete (hence locally compact and metrizable) topological group G homeomorphic to a closed subspace of a separable Tychonoff space is not necessarily separable. Indeed, it suffices to consider the Niemytzki plane which contains a discrete copy of the real numbers, the X-axis. Therefore Theorem 8 and Corollaries 3 and 4 would not be valid if the group G were assumed to be a subspace of a separable Hausdorff (or even Tychonoff) space rather than a subgroup of a separable topological group.

(2) The separable connected pro-Lie group $G = \mathbb{R}^{c}$ contains a closed non-separable subgroup. To see this, we consider the closed subgroup $\mathbb{Z}^{c}$ of G. By a theorem of Uspenskij [28], the group $\mathbb{Z}^{c}$ contains a subgroup H of uncountable cellularity. The closure of H in G, say, K is a closed non-separable subgroup of G. Note that the group K cannot be almost connected. (See the next subsection for discussion of almost connected groups.)

(3) A natural question is whether a connected metrizable group must be separable if it is a subspace of a separable Hausdorff (or regular) space. Again the answer is 'No'. Indeed, consider an arbitrary connected metrizable group G of weight c. For example, one can take $G = C(X)$, the Banach space of continuous real-valued functions on a compact space X satisfying $w(X) = c$, endowed with the sup-norm topology. Since $w(G) = c$, the space G is homeomorphic to a subspace of the Tychonoff cube I^{c}, where $I = [0, 1]$ is the closed unit interval. Thus G embeds as a subspace in a separable regular space, but both the density and weight of G are equal to c.

3.3. Separability of Pro-Lie Groups

Early this century Hofmann and Morris, [17,29], introduced the class of *pro-Lie groups*, which consists of projective limits of finite-dimensional Lie groups and proved that it contains all compact groups, all locally compact abelian groups, and all connected locally compact groups and is closed under the formation of products and closed subgroups. They defined a topological group G to be *almost connected* if the quotient group of G by the connected component of its identity is compact [17]. Of course all compact groups, all connected topological groups and all finite or infinite products of a set of topological groups, each factor of which is either a connected topological group or a compact group, are almost connected.

Below we consider topological groups which are *homeomorphic* to a subspace of a separable *Hausdorff* space.

Theorem 9 ([4]). *Let G be an almost connected pro-Lie group. If G is homeomorphic to a subspace of a separable Hausdorff space, then G is separable.*

This result can be strengthened as follows.

Theorem 10 ([4]). *Let G be an ω-narrow topological group which contains a closed subgroup N such that N is an almost connected pro-Lie group and the quotient space G/N is locally compact. If G is homeomorphic to a subspace of a separable Hausdorff space, then G is separable.*

Unlike the case of almost connected pro-Lie groups, closed subgroups of separable prodiscrete abelian groups can fail to be separable.

Proposition 6 ([4]). *Closed subgroups of separable prodiscrete abelian groups need not be separable.*

3.4. Closed Topologically Isomorphic Embeddings into Separable Topological Groups

Theorems 11 and 12 given below show that there is a wealth of separable pseudocompact topological (abelian) groups with closed non-separable subgroups.

Theorem 11 ([4]). *Every precompact topological group of weight $\leq$ c is topologically isomorphic to a closed subgroup of a separable, connected, pseudocompact group H of weight $\leq$ c.*

Here is the abelian version.

Theorem 12 ([4]). *Every precompact abelian group of weight $\leq$ c is topologically isomorphic to a closed subgroup of a separable, connected, pseudocompact abelian group H of weight $\leq$ c.*

Theorem 13 ([4]). *Under the Continuum Hypothesis CH, there exists a separable countably compact abelian topological group G which contains a closed non-separable subgroup.*

Problem 4. *Does there exist in ZFC a separable countably compact topological group which contains a non-separable closed subgroup?*

4. Products of Separable Topological Groups/Locally Convex Spaces

4.1. Strongly Separable Topological Groups

Let us say that a topological group G is *strongly separable* (briefly, *S-separable*) if for any topological group H such that every closed subgroup of H is separable, the product $G \times H$ has each of its closed subgroups separable. What are the *S*-separable groups?

One of the main statements is the following result which can be reformulated by saying that every separable compact group is *S*-separable.

Theorem 14 ([1]). *Let G be a separable compact group and H be a topological group in which all closed subgroups are separable. Then all closed subgroups of the product $G \times H$ are separable.*

It is not clear to what extent one can generalize Theorem 14 by weakening the compactness assumption on G. However by Theorem 17 some additional conditions on the groups G and/or H have to be imposed.

The proof of Theorem 14 relies on the fact that for a compact factor G the projection $G \times H$ onto H is a closed mapping. In the next proposition we present another situation when the projection $G \times H \rightarrow H$ turns out to be a closed mapping.

Proposition 7 ([1]). *Let G be a countably compact topological group and H a separable metrizable topological group. If all closed subgroups of G are separable, then all closed subgroups of the product $G \times H$ are separable.*

Every countable group is *S*-separable. The theorem below unites both the compact and countable classes of groups.

Theorem 15 ([1]). *A topological group G is S-separable provided it contains a separable compact subgroup K such that the quotient space G/K is countable.*

Proposition 8 ([1]). *The class of S-separable groups is closed under the operations:*

(1) *finite products;*
(2) *taking closed subgroups;*

(3)   *taking continuous homomorphic images.*

4.2. Product of Two Separable Precompact/Pseudocompact Groups

Theorem 16 ([1]). *Assume that $2^{\omega_1} = \mathfrak{c}$. Then there exist pseudocompact abelian topological groups G and H such that all closed subgroups of G and H are separable, but the product $G \times H$ contains a closed non-separable σ-compact subgroup.*

Recently Zhiqiang Xiao, Sánchez and Tkachenko [6] presented the first example in ZFC of precompact (but not necessarily pseudocompact) abelian topological groups G and H with the similar properties.

Theorem 17 ([6]). *There exist precompact abelian topological groups G and H such that all closed subgroups of G and H are separable, but the product $G \times H$ contains a closed non-separable subgroup.*

Also the authors of [6] improved upon Theorem 16 by constructing, under the assumption of $2^{\omega_1} = \mathfrak{c}$, a pseudocompact abelian topological group G such that every closed subgroup of G is separable, but the square $G \times G$ contains a closed non-separable σ-compact subgroup.

Finally, the following result is obtained under the assumption of Martin's Axiom and negation of the Continuum Hypothesis $MA\&\neg CH$.

Theorem 18 ([6]). *Assume $MA\&\neg CH$. Then there exist countably compact Boolean topological groups G and H such that all closed subgroups of G and H are separable, but the product $G \times H$ contains a closed non-separable subgroup.*

4.3. Product of Two Separable Pseudocomplete Locally Convex Spaces

Proposition 9 ([1]). *Let K be a finite-dimensional Banach space and let L be a topological vector space in which all closed vector subspaces are separable. Then all closed vector subspaces of the product $K \times L$ are separable.*

Remark 3. It is not known whether Proposition 9 remains valid for an arbitrary separable Banach space K.

Theorem 19 ([1]). *Assume that $2^{\omega_1} = \mathfrak{c}$. Then there exist pseudocomplete locally convex spaces K and L such that all closed vector subspaces of K and L are separable, but the product $K \times L$ contains a closed non-separable σ-compact vector subspace.*

4.4. Product of Continuum Many Separable Locally Convex Spaces

The classical Hewitt–Marczewski–Pondiczery Theorem 2 implies that the product of no more than $\mathfrak{c}$ separable topological spaces is separable. Domański [30] gave an example of a non-separable complete locally convex space which can be embedded as a closed vector subspace of a product of $\mathfrak{c}$ copies of the Banach space c_0. Later he extended this result in show that every product of $\mathfrak{c}$ copies of any infinite-dimensional Banach space has non-separable closed vector subspaces [31]. In fact, Domański proved in [31] that if E_i, $i \in I$, with card$(I) = \mathfrak{c}$ are separable topological vector spaces whose completions are not q-minimal, then the product $\prod_{i \in I} E_i$ has a non-separable closed vector subspace. (A topological vector space E is called *q-minimal* if it and all its quotient spaces are minimal, while E is called *minimal* if it does not admit a strictly weaker Hausdorff topological vector space topology).

Similarly to the variety of topological groups defined earlier, a non-empty class Ω of locally convex spaces is said to be a *variety* [27,32–34] if it is closed under the operations of taking subspaces, quotient spaces, (arbitrary) cartesian products and isomorphic images. Let $\mathcal{C}$ be a class of locally convex spaces, denote by $\mathfrak{V}(\mathcal{C})$ the intersection of all varieties containing $\mathcal{C}$. Then $\mathfrak{V}(\mathcal{C})$ is said to be the *variety generated by* $\mathcal{C}$. We repeat that if $\mathcal{C}$ consists of a single object E, then $\mathfrak{V}(\mathcal{C})$ is written as $\mathfrak{V}(E)$.

Theorem 20 ([3]). *Let I be an index set and E_i a locally convex space for each $i \in I$. If at least $\mathfrak{c}$ of the E_i are not in $\mathfrak{V}(\mathbb{R})$, or equivalently do not have the weak topology, then the product $\prod_{i\in I} E_i$ has a non-separable closed vector subspace.*

Theorem 21 ([3]). *Let I be any index set and each E_i, $i \in I$, a separable locally convex space. If X is a vector subspace of $\prod_{i\in I} E_i$ with Y a closed vector subspace of X such that either*
    (a) *Y is metrizable and X/Y is separable, or*
    (b) *Y is separable and X/Y is metrizable,*
then X is separable.

Let $C_p(X, E)$ denote the space of all continuous E-valued functions on X endowed with the pointwise convergence topology, where E is a locally convex space. The space $C_p(X, E)$ is a vector subspace of E^X endowed with the product topology.

Corollary 6 ([3]). *Let X be a Tychonoff space. If E is a separable locally convex space, then every metrizable vector subspace of $C_p(X, E)$ is separable.*

Proposition 10 ([3]). *Let X be a Tychonoff space such that every closed vector subspace of $C_p(X)$ is separable. Then every closed subset F of X is a G_δ-set, that is $F = \bigcap_{i=1}^{\infty} U_i$, where each U_i is open in X.*

Example 2 ([3]). Let $\mathbb{M}$ denote the Michael line. Then $C_p(\mathbb{M})$ is a separable locally convex space containing a non-separable closed vector subspace.

4.5. Open Problems

Problem 5 ([1]). *Find the frontiers of the class of S-separable topological groups:*

(a)   *Is every separable locally compact group S-separable?*
(b)   *Is the abelian topological group $\mathbb{R}$ of all real numbers with the Euclidean topology S-separable? Does there exist a separable metrizable group which is not S-separable?*
(c)   *Is the free topological group on the closed unit interval $[0, 1]$ S-separable?*

The following problem arises in an attempt to generalize Proposition 7:

Problem 6 ([1]). *Let G be a countably compact topological group such that all closed subgroups of G are separable, and H a topological group with a countable network. Are the closed subgroups of $G \times H$ separable?*

Denote by $\mathfrak{S}$ the smallest class of topological groups which is generated by all compact separable groups and all countable groups and is closed under the operations listed in (1)–(3) of Proposition 8. It is not difficult to verify that if $G \in \mathfrak{S}$, then G contains a compact separable subgroup K such that the quotient space G/K is countable. In the next problem we ask if this property characterizes the groups from $\mathfrak{S}$:

Problem 7 ([1]). *Does a topological group G belong to the class $\mathfrak{S}$ if and only if G contains a compact separable subgroup K such that the quotient space G/K is countable?*

Problem 8 ([1]). *Are Theorems 16 and 19 valid in ZFC alone?*

Problem 9 ([3]). *Characterize those Tychonoff spaces X such that all closed vector subspaces of $C_p(X)$ are separable.*

5. The Separable Quotient Problem for General Topological Groups

Let us begin this section with a famous unsolved problem in Banach space theory. The Separable Quotient Problem for Banach Spaces has its roots in the 1930s and is due to Stefan Banach and Stanisław Mazur.

Problem 10. *(Separable Quotient Problem for Banach Spaces) Does every infinite-dimensional Banach space have a quotient Banach space which is separable and infinite-dimensional?*

In the literature many special cases of the Separable Quotient Problem for Banach Spaces have been proved, for instance:

- Every infinite-dimensional reflexive Banach space has a separable infinite-dimensional quotient Banach space (Pełczyński, 1964).
- Every Banach space $C(K)$, where K is a compact space, has a separable infinite-dimensional quotient Banach space (Rosenthal, 1969; Lacey, 1972).
- Every Banach dual of any infinite-dimensional Banach space, E^*, has a separable infinite-dimensional quotient Banach space (Argyros, Dodos, Kanellopoulos, 2008).

However, the general Problem 10 remains unsolved.

Turning to locally convex spaces one can state the analogous problem.

Question 1. *(Separable Quotient Problem for Locally Convex Spaces) Does every infinite-dimensional locally convex space have a quotient locally convex space which is* separable *and infinite-dimensional?*

- Every infinite-dimensional Fréchet space which is non-normable has the separable metrizable topological vector space $\mathbb{R}^\omega$ as a quotient space (Eidelheit, 1936).

Please note that there are many other partial positive solutions in the literature to Problem 1 (see [35]). However, Kąkol, Saxon and Todd [36] answered Question 1 in the negative. Recall that a *barrel* in a topological vector space is a convex, balanced, absorbing and closed set. A Hausdorff topological vector space E is called barreled if every barrel in E is a neighborhood of the zero element.

Theorem 22 ([36]). *There exists an infinite-dimensional barreled locally convex space without any quotient space which is an infinite-dimensional separable locally convex space.*

Now we formulate various natural versions of the Separable Quotient Problem(s) for Topological Groups. Unless explicitly stated otherwise the results presented in this section are from the paper [5].

Problem 11. *(Separable Quotient Problem for Topological Groups) Does every non-totally disconnected topological group have a quotient group which is a non-trivial separable topological group?*

Problem 12. *(Separable Infinite Quotient Problem for Topological Groups) Does every non-totally disconnected topological group have a quotient group which is an infinite separable topological group?*

Problem 13. *(Separable Metrizable Quotient Problem for Topological Groups) Does every non-totally disconnected topological group have a quotient group which is a non-trivial separable metrizable topological group?*

Problem 14. *(Separable Infinite Metrizable Quotient Problem for Topological Groups) Does every non-totally disconnected topological group have a quotient group which is an infinite separable metrizable topological group?*

It is natural to consider these questions for various prominent classes of topological groups such as Banach spaces, locally convex spaces, compact groups, locally compact groups, pro-Lie groups, pseudocompact groups, and precompact groups. The paper [7] provides an interesting solution for Banach spaces.

Theorem 23 ([7]). *Let E be a locally convex space (over $\mathbb{R}$ or $\mathbb{C}$). If E has a subspace which is an infinite-dimensional Fréchet space, then E has the (infinite separable metrizable) tubby torus group $\mathbb{T}^{\omega}$ as a quotient group.*

Corollary 7 ([7]). *Every infinite-dimensional Fréchet space, and in particular every infinite-dimensional Banach space, has the (infinite separable metrizable) tubby torus group $\mathbb{T}^{\omega}$ as a quotient group.*

We denote by φ the complete countable infinite-dimensional locally convex space which is the strong dual of the locally convex space $\mathbb{R}^{\omega}$.

Remark 4 ([7]). *There is no continuous surjective homomorphism of the separable locally convex space φ onto the tubby torus $\mathbb{T}^{\omega}$.*

Corollary 7 suggests the following unsolved problem, a negative answer for which would immediately yield a negative answer to the Banach-Mazur Separable Quotient Problem for Banach Spaces, Problem 10.

Problem 15. *Does every infinite-dimensional Banach space have a quotient group which is homeomorphic to $\mathbb{R}^{\omega}$?*

A topological group G is said to be a *SIN-group* if every neighborhood of the identity of G contains a neighborhood of the identity which is invariant under all inner automorphisms. Of course every abelian topological group and every compact group is a SIN-group.

As we noted in Section 2 the cardinality of any regular separable topological space is not greater than $2^{\mathfrak{c}}$. This makes the following statement interesting.

Theorem 24 ([37]). *Let G be an abelian topological group or more generally a SIN-group. If G has a quotient group which is separable, then it also has a quotient group of cardinality not greater than $\mathfrak{c}$.*

We now consider several natural questions which are special cases of Problems 11, 12, 13, and 14.

Question 2. *(Separable Quotient Problem for Locally Compact Abelian Groups) Does every infinite locally compact abelian group have a separable quotient group which is (i) non-trivial; (ii) infinite; (iii) metrizable; (iv) infinite metrizable?*

The non-abelian version of Question 2 is:

Question 3. *(Separable Quotient Problem for Locally Compact Groups) Does every non-totally disconnected locally compact group have a separable quotient group which is (i) non-trivial; (ii) infinite; (iii) metrizable; (iv) infinite metrizable?*

As a special case of Question 3 we have:

Question 4. *(Separable Quotient Problem for Compact Groups) Does every infinite compact group have a separable quotient group which is (i) non-trivial; (ii) infinite; (iii) metrizable; (iv) infinite metrizable?*

5.1. Locally Compact Groups and Pro-Lie Groups

In this subsection we present a positive answer to each of Question 2 (i), (ii), (iii), and (iv) and Question 4 (i), (ii), (iii), and (iv), and a partial answer to Question 3. Satisfying results have been proved for pro-Lie groups. Stronger structural results have been obtained for compact abelian groups, connected compact groups, and totally disconnected compact groups.

Theorem 25. *Every non-separable compact abelian group G has a quotient group Q which is a countably infinite product of non-trivial compact finite-dimensional Lie groups. The quotient group, Q, is therefore an infinite separable metrizable group.*

Theorem 26. *Every non-separable connected compact group, G, has a quotient group, Q, which is a countably infinite product of non-trivial compact finite-dimensional Lie groups. The quotient group, Q, is therefore an infinite separable metrizable group.*

Remark 5. No discrete group has a quotient group which is a countably infinite product of non-trivial topological groups since every quotient of a discrete group is evidently discrete. In particular then, a locally compact abelian group need not have a quotient group which is a countably infinite product of non-trivial topological groups.

Theorem 27. *Every non-separable connected locally compact abelian group, G, has a quotient group, Q, which is a countably infinite product of non-trivial compact finite-dimensional Lie groups. The quotient group, Q, is therefore an infinite separable metrizable group.*

Theorem 28. *Every infinite totally disconnected compact group G has a quotient group, Q, which is homeomorphic to a countably infinite product of finite discrete topological groups. The quotient group, Q, is thus homeomorphic to the Cantor space and therefore is an infinite separable metrizable group.*

Below a positive answer to Question 4 (i), (ii), (iii), and (iv) is presented.

Theorem 29. (Separable Quotient Theorem for Compact Groups) *Let G be an infinite compact group. Then G has a quotient group which is an infinite separable metrizable (compact) group.*

With the help of Theorem 29 a positive answer to Question 2 (i), (ii), (iii), and (iv) is given.

Theorem 30. (Separable Quotient Theorem for Locally Compact Abelian Groups) *Let G be an infinite locally compact abelian group. Then G has a quotient group which is an infinite separable metrizable group.*

Recall that a *proto-Lie group* is defined in ([17], Definition 3.25) to be a topological group G for which every neighborhood of the identity contains a closed normal subgroup N such that the quotient group G/N is a Lie group. If G is also a complete topological group, then it is said to be a *pro-Lie group*. If G is a proto-Lie group (respectively, pro-Lie group) with all the quotient Lie groups G/N discrete then G is said to be *protodiscrete* (respectively, *prodiscrete*). It is immediately clear that if G is a proto-Lie group which is not a Lie group, then it is not topologically simple.

Theorem 31. (Separable Quotient Theorem for Proto-Lie Groups) *Let G be an infinite proto-Lie group which is not protodiscrete; that is, G is not totally disconnected. Then G has a quotient group which is an infinite separable metrizable (Lie) group.*

Theorem 32. (Separable Quotient Theorem for σ-compact Pro-Lie groups) *Let G be an infinite σ-compact pro-Lie group. Then G has a quotient group which is an infinite separable metrizable group.*

Another significant generalization of Theorem 30 is Theorem 33.

Theorem 33. *(Separable Quotient Theorem for Abelian Pro-Lie groups) Let G be an infinite abelian pro-Lie group. Then G has a quotient group which is an infinite separable metrizable group.*

The next theorem, which generalizes Theorem 29, provides a partial but significant answer to Question 3.

Theorem 34. *(Separable Quotient Theorem for σ-compact Locally Compact Groups) Every infinite σ-compact locally compact group has a quotient group which is an infinite separable metrizable group.*

Corollary 8. *(Separable Quotient Theorem for Almost Connected Locally Compact Groups) Every infinite almost connected locally compact group has a quotient group which is an infinite separable metrizable group.*

5.2. σ-Compact Groups, Lindelöf Σ-Groups and Pseudocompact Groups

Recall that the class of Lindelöf Σ-groups contains all σ-compact and all separable metrizable topological groups, and is closed with respect to countable products, closed subgroups, and continuous homomorphic images (see [10], Section 5.3).

Proposition 11. *(Separable Quotient Theorem for Lindelöf Σ-groups) Let G be an infinite Lindelöf Σ-group. Then G has a quotient group which is infinite and separable. Indeed, the topology of G is initial with respect to the family of quotient homomorphisms onto infinite groups with a countable network.*

Since every σ-compact topological group is evidently a Lindelöf Σ-group, the next result is immediate from Proposition 11.

Corollary 9. *(Separable Quotient Theorem for σ-compact Groups) Let G be an infinite σ-compact topological group. Then G has a quotient group which is infinite and separable. Indeed, the topology of G is initial with respect to the family of quotient homomorphisms of the group onto infinite groups with a countable network.*

Regarding Question 14, one might reasonably ask: If the topological group G has a quotient group which is infinite and separable, does G necessarily have a quotient group which is infinite, separable and metrizable? This question is answered negatively in the next Proposition 12.

Proposition 12. *There exists a countably infinite precompact abelian group H such that every quotient group of H is either trivial or non-metrizable.*

We now consider pseudocompact groups.

Theorem 35. *(Separable Quotient Theorem for Pseudocompact Groups) The topology of every infinite pseudocompact topological group, G, is initial with respect to the family of quotient homomorphisms onto infinite compact metrizable groups. In particular, G has a quotient group which is infinite separable compact and metrizable.*

5.3. A Precompact Topological Group Which Does Not Admit Separable Quotient Group

In this section, we show that Theorem 35 cannot be extended to precompact topological groups, even in the weak form of the existence of nontrivial separable quotients.

Theorem 36. *There exists an uncountable dense subgroup G of the compact group $\mathbb{T}^{\mathfrak{c}}$ satisfying $\dim G = 0$ such that every countable subgroup of G is closed in G and every uncountable subgroup of G is dense in G. Hence every quotient group of G is either trivial or non-separable.*

In fact, every power of the group G in Theorem 36 does not have non-trivial separable quotients.

Theorem 37. *Let $G \subset \mathbb{T}^{\mathfrak{c}}$ be the group constructed in Theorem 36 and let $\tau \geq 1$ be a cardinal number. Then every quotient group of G^{τ} is either trivial or non-separable.*

A topological group is said to be a G_{σ}-*group* if it has a dense subgroup H which is the union of a strictly increasing sequence of closed topological subgroups. Finally, we show that there exists a G_{σ}-*group* without a nontrivial separable quotient group.

Theorem 38. *For every cardinal $\tau \geq \mathfrak{c}$, there exists a precompact topological abelian group H satisfying $\dim G = 0$ and with the following properties:*

(a) $w(H) = \tau$;
(b) $H = \bigcup_{n \in \omega} H_n$, *where $H_0 \subset H_1 \subset H_2 \subset \cdots$ are proper closed subgroups of H;*
(c) *every quotient group of H is either trivial or non-separable.*

The class of $\mathbb{R}$-*factorizable* groups (see [10], Chapter 8) contains all pseudocompact groups as well as σ-compact groups. We note that by Theorem 8.1.9 of [10], a locally compact group is $\mathbb{R}$-factorizable if and only if it is σ-compact. Since the group G in Theorem 36 is precompact, it is $\mathbb{R}$-factorizable according to ([10], Corollary 8.1.17).

Corollary 10. *There exist infinite $\mathbb{R}$-factorizable groups without non-trivial separable or metrizable quotients.*

5.4. Open Problems

We note that Question 3 formulated earlier in Section 5 has not been fully answered, so we state it now as an unsolved problem.

Problem 16. (Separable Quotient Problem for Locally Compact Groups) *Does every infinite non-totally disconnected locally compact group have a separable quotient group which is (i) non-trivial; (ii) infinite; (iii) metrizable; (iv) infinite metrizable?*

Recall that an abelian topological group G is called a *reflexive topological group* if the natural map of G into its second dual group is a topological group isomorphism. The Pontryagin van-Kampen Theorem [38] says that every locally compact abelian group is reflexive. It is also known that every complete metrizable locally convex space, in particular every Banach space, is a reflexive topological group ([39], Proposition 15.2). Therefore the following unsolved problem arises naturally.

Problem 17. (Separable Quotient Problem for Reflexive Topological Groups) *Does every infinite reflexive abelian topological group, G, have a separable quotient group which is (i) non-trivial; (ii) infinite; (iii) metrizable; (iv) infinite metrizable?*

Problem 18. *Does there exist a precompact abelian group G as in Theorem 36 which has one of the following additional properties:*

(a) *G is connected;*
(b) *G is Baire;*
(c) *G is reflexive?*

6. Quotient Groups of Free Topological Groups

Let, as usual, $F(X)$ and $A(X)$ denote the free topological group and the free abelian topological group of a Tychonoff space X, respectively. $A(X)$ is a natural quotient group of $F(X)$, and for every X there is a quotient mapping from $A(X)$ onto the group of integers $\mathbb{Z}$ (see [10], Chapter 7).

6.1. Free Topological Groups Which Admit Second Countable Quotient Groups

A space X is called *ω-bounded* if the closure of every countable subset of X is compact. Clearly, every compact space is ω-bounded, while every ω-bounded space is countably compact.

Proposition 13 ([2]). *Let X be a non-scattered Tychonoff space. If X has one of the following properties (a) or (b), then both $A(X)$ and $F(X)$ admit an open continuous homomorphism onto the circle group $\mathbb{T}$:*

(a) *X is normal and countably compact;*
(b) *X is ω-bounded.*

Proposition 14 ([2]). *Let X be a scattered ω-bounded Tychonoff space. Then every quotient group of $F(X)$ and $A(X)$ is either discrete and finitely generated (hence, countable) or non-metrizable.*

Theorem 39 ([2]). *Let X be an ω-bounded Tychonoff space. Then the following conditions are equivalent:*

(a) *Every metrizable quotient group of $F(X)$ is discrete and finitely generated.*
(b) *Every metrizable quotient group of $F(X)$ is finitely generated.*
(c) *Every metrizable quotient group of $F(X)$ is countable.*
(d) *X is scattered.*

Corollary 11 ([2]). *Let X be either the compact space of ordinals $[0, \alpha]$ with the order topology or the one-point compactification of an arbitrary discrete space. Then every metrizable quotient group of $F(X)$ or $A(X)$ is discrete and finitely generated.*

It turns out that free topological groups on non-pseudocompact zero-dimensional spaces do have non-trivial metrizable quotient groups:

Proposition 15 ([2]). *Let X be a non-pseudocompact zero-dimensional space. Then the groups $F(X)$ and $A(X)$ admit an open continuous homomorphism onto the (countably infinite separable metrizable) discrete group $A(\mathbb{Z})$.*

6.2. Free Topological Groups Which Admit Quotient Groups with a Countable Network

Proposition 16 ([2]). *Let X be a locally compact or pseudocompact space. Then the groups $A(X)$ and $F(X)$ admit an open continuous homomorphism onto $A(S)$, where S is an infinite compact subspace of the closed unit interval $[0, 1]$, hence $A(S)$ has a countable network.*

Proposition 17 ([2]). *Let X be a Lindelöf Σ-space (in particular, σ-compact space). Then the groups $A(X)$ and $F(X)$ admit an open continuous homomorphism onto $A(Y)$, where Y has a countable network, hence $A(Y)$ also has a countable network.*

6.3. Free Topological Groups Which Admit Separable Quotient Groups

Theorem 40 ([2]). *Let X be a Tychonoff space satisfying the following conditions:*

(1) *the closure of every countable subset of X is countable and compact;*
(2) *every countable compact subset of X is a retract of X.*

Then every separable quotient group of $F(X)$ is countable.

Corollary 12 ([2]). *Let X be either the space of ordinals $[0, \alpha)$ with the order topology or the one-point compactification of an arbitrary discrete space. Then every separable quotient group of $F(X)$ or $A(X)$ is countable.*

6.4. Open Problems

Similarly to Problems 1 and 2 we can ask the following related questions.

Problem 19 ([2]). *Does there exist an open continuous homomorphism of $A(X)$ onto $A(S)$, where X is an arbitrary Tychonoff space and S is an infinite subspace of the closed unit interval $[0, 1]$?*

Furthermore, a more particular question below is open:

Problem 20 ([2]). *Does there exist an open continuous homomorphism of $A(X)$ onto $A(Y)$, where X is an arbitrary Lindelöf space and Y is an infinite space which has a countable network?*

7. Separable Group Topologies for Abelian Groups

Which abelian groups G admit a separable Hausdorff group topology? To answer the question, the author of [8] considers three different cases:
Case 1. There is an element $x \in G$ of infinite order;
Case 2. G is a bounded torsion group;
Case 3. G is an unbounded torsion group.

Theorem 41 ([8]). *Let G be an abelian group with $|G| \leq 2^c$. Then G admits a separable, precompact, Hausdorff group topology.*

Remark 6. The "abelian" condition in Theorem 41 cannot be deleted as Shelah [40] proved that there exist non-abelian groups which admit no non-discrete Hausdorff topological group topology. Morris and Obraztsov ([41], Theorem L) produce an uncountable number of countably infinite groups each of which admits no non-discrete Hausdorff topological group topology; more precisely, they identify a continuum of pairwise non-isomorphic infinite groups, $\{G_i : i \in I\}$, of exponent p^2, for any sufficiently large prime p, where each proper subgroup of G_i is cyclic and each G_i does not admit any non-discrete Hausdorff topological group topology. It is also proved in [41] that there exist non-discrete Hausdorff topological groups G of each cardinality $\aleph_n$ with no proper subgroup of the same cardinality as G. Such groups obviously have no quotient group of smaller cardinality than that of G.

Remark 7. Yves Cornulier noticed that every abelian group G of cardinality $|G| \leq 2^c$ embeds as a subgroup of $(\mathbb{Q} \times \mathbb{Q}/\mathbb{Z})^c$, which is a separable Hausdorff topological group. Nevertheless, this remark does not prove Theorem 41, because even a closed subgroup of a separable group, with the induced topology, need not to be separable.

Yves Cornulier also kindly informed us that in sharp contrast to the abelian case, for every uncountable cardinal τ, there exists a 2-step nilpotent group of cardinality τ that has no Hausdorff separable group topology. So any reasonable potential generalization of Theorem 41 fails.

Author Contributions: This survey is the joint effort of the authors, who contributed equally.

Funding: This research received no external funding.

Conflicts of Interest: The authors declare no conflict of interest.

References

1. Leiderman, A.G.; Tkachenko, M.G. Products of topological groups in which all closed subgroups are separable. *Topol. Appl.* **2018**, *241*, 89–101. [CrossRef]
2. Leiderman, A.G.; Tkachenko, M.G. Quotients of free topological groups. preprint, 2018.
3. Kąkol, J.; Leiderman, A.G.; Morris, S.A. Nonseparable closed vector subspaces of separable topological vector spaces. *Monatsh. Math.* **2017**, *182*, 39–47; Erratum in **2017**, *182*, 49–50. [CrossRef]
4. Leiderman, A.G.; Morris, S.A.; Tkachenko, M.G. Density character of subgroups of topological groups. *Trans. Am. Math. Soc.* **2017**, *369*, 5645–5664. [CrossRef]
5. Leiderman, A.G.; Morris, S.A.; Tkachenko, M.G. The separable quotient problem for topological groups. *arXiv* **2017**, arXiv:1707.09546.
6. Xiao, Z.; Sánchez, I.; Tkachenko, M. Topological groups whose closed subgroups are separable and the product operation. *Topol. Appl.* **2019**, to appear.
7. Gabriyelyan, S.S.; Morris, S.A. A topological group observation on the Banach-Mazur separable quotient problem. *Topol. Appl.* **2019**, to appear.
8. López, L.F.M. Continuous isomorphisms onto separable groups. *Appl. Gen. Topol.* **2012**, *13*, 135–150.
9. Engelking, R. *General Topology*; Heldermann Verlag: Berlin, Germany, 1989.
10. Arhangel'skii, A.V.; Tkachenko, M.G. *Topological Groups and Related Structures*; Atlantis Series in Mathematics, Volume I; Atlantis Press and World Scientific: Paris, France; Amsterdam, The Netherlands, 2008.
11. Hodel, R.E. Cardinal Functions I, Chapter 1. In *Handbook of Set-Theoretic Topology*; Kunen, K., Vaughan, J.E., Eds.; North Holland: Amsterdam, The Netherlands; New York, NY, USA; Oxford, UK, 1984.
12. Hofmann, K.H.; Morris, S.A. *The Structure of Compact Groups*, 3rd ed.; Walter de Gruyter Gmbh: Berlin, Germany; Boston, MA, USA, 2013.
13. Engelking, R. Cartesian products and dyadic spaces. *Fundam. Math.* **1965**, *57*, 287–304. [CrossRef]
14. Karnik, S.M.; Willard, S. Natural covers and *r*-quotient mappings. *Can. Math. Bull.* **1982**, *25*, 456–462. [CrossRef]
15. Fujita, H.; Shakhmatov, D. Topological groups with dense compactly generated subgroups. *Appl. Gen. Topol.* **2002**, *3*, 85–89. [CrossRef]
16. Guran, I.I. On topological groups close to being Lindelöf. *Sov. Math. Dokl.* **1981**, *23*, 173–175.
17. Hofmann, K.H.; Morris, S.A. *The Lie Theory of Connected Pro-Lie Groups*; European Mathematical Society: Zurich, Switzerland, 2007.
18. Itzkowitz, G.L. On the density character of compact topological groups. *Fundam. Math.* **1972**, *75*, 201–203. [CrossRef]
19. Comfort, W.W.; Itzkowitz, G.L. Density character in topological groups. *Math. Ann.* **1977**, *226*, 223–227. [CrossRef]
20. Vidossich, G. Characterization of separability for LF-spaces. *Annales de L'institut Fourier* **1968**, *18*, 87–90. [CrossRef]
21. Lohman, R.H.; Stiles, W.J. On separability in linear topological spaces. *Proc. Am. Math. Soc.* **1974**, *42*, 236–237. [CrossRef]
22. Morris, S.A. Varieties of topological groups. *Bull. Aust. Math. Soc.* **1969**, *1*, 145–160. [CrossRef]
23. Morris, S.A. Varieties of topological groups II. *Bull. Aust. Math. Soc.* **1970**, *2*, 1–13. [CrossRef]
24. Morris, S.A. Varieties of topological groups III. *Bull. Aust. Math. Soc.* **1970**, *2*, 165–178. [CrossRef]
25. Morris, S.A. Varieties of topological groups a survey. *Colloq. Math.* **1982**, *2*, 147–165. [CrossRef]
26. Brooks, M.S.; Morris, S.A.; Saxon, S.A. Generating varieties of topological groups. *Proc. Edinb. Math. Soc.* **1973**, *18*, 191–197. [CrossRef]
27. Kopperman, R.; Mislove, M.W.; Morris, S.A.; Nickolas, P.; Pestov, V.; Svetlichny, S. Limit laws for wide varieties of topological groups. *Houst. J. Math.* **1996**, *22*, 307–328.
28. Uspenskij, V.V. On the Suslin number of subgroups of products of countable groups. *Acta Universitatis Carolinae—Mathematica et Physica* **1995**, *36*, 85–87.
29. Hofmann, K.H.; Morris, S.A. Pro-Lie groups: A survey with open problems. *Axioms* **2015**, *4*, 294–312. [CrossRef]
30. Domański, P. On the separable topological vector spaces. *Funct. Approx. Comment. Math.* **1984**, *14*, 117–122.

31. Domański, P. Nonseparable closed subspaces in separable products of topological vector spaces, and *q*-minimality. *Arch. Math.* **1983**, *41*, 270–275. [CrossRef]
32. Diestel, J.; Morris, S.A.; Saxon, S.A. Varieties of locally convex topological vector spaces. *Bull. Am. Math. Soc.* **1971**, *77*, 799–803. [CrossRef]
33. Diestel, J.; Morris, S.A.; Saxon, S.A. Varieties of linear topological spaces. *Trans. Am. Math. Soc.* **1972**, *172*, 207–230. [CrossRef]
34. Morris, S.A.; Diestel, J. Remarks on varieties of locally convex linear topological spaces. *J. Lond. Math. Soc.* **1974**, *8*, 271–278. [CrossRef]
35. Kąkol, J.; Saxon, S.A. Separable quotients in $C_c(X)$, $C_p(X)$, and their duals. *Proc. Am. Math. Soc.* **2017**, *145*, 3829–3841. [CrossRef]
36. Kąkol, J.; Saxon, S.A.; Todd, A. Barrelled spaces with(out) separable quotients. *Bull. Aust. Math. Soc.* **2014**, *90*, 295–303. [CrossRef]
37. Hofmann, K.H.; Morris, S.A.; Nickolas, P.; Pestov, V. Small large subgroups of a topological group. *Note Math.* **1997**, *4*, 161–165.
38. Morris, S.A. *Pontryagin Duality and the Structure of Locally Compact Abelian Groups*; Cambridge University Press: Cambridge, UK, 1977; ISBN 0521215439.
39. Banaszczyk, W. *Additive Subgroups of Topological Vector Spaces*; Springer: Berlin/Heidelberg, Germany; New York, NY, USA, 1977; ISBN 0387539174.
40. Shelah, S. On a problem of Kurosh, Jonsson groups, and applications. In *Word Problems Vol II*; Amsterdam, The Netherlands, 1980; pp. 373–394.
41. Morris, S.A.; Obraztsov, V.N. Non-discrete topological groups with many discrete subgroups. *Topol. Appl.* **1998**, *84*, 105–120. [CrossRef]

MDPI

Article

Categorically Closed Topological Groups

Taras Banakh [1,2]

[1] Faculty of Mechanics and Mathematics, Ivan Franko National University, 79000 Lviv, Ukraine;
t.o.banakh@gmail.com
[2] Institute of Mathematics, Jan Kochanowski University, 25-001 Kielce, Poland

Received: 30 June 2017; Accepted: 27 July 2017; Published: 30 July 2017

Abstract: Let $\vec{\mathcal{C}}$ be a category whose objects are semigroups with topology and morphisms are closed semigroup relations, in particular, continuous homomorphisms. An object X of the category $\vec{\mathcal{C}}$ is called $\vec{\mathcal{C}}$-*closed* if for each morphism $\Phi \subset X \times Y$ in the category $\vec{\mathcal{C}}$ the image $\Phi(X) = \{y \in Y : \exists x \in X \ (x, y) \in \Phi\}$ is closed in Y. In the paper we survey existing and new results on topological groups, which are $\vec{\mathcal{C}}$-closed for various categories $\vec{\mathcal{C}}$ of topologized semigroups.

Keywords: topological group; paratopological group; topological semigroup; absolutely closed topological group; topological group of compact exponent

---

1. Introduction and Survey of Main Results

In this paper, we recognize topological groups which are $\vec{\mathcal{C}}$-closed for some categories $\vec{\mathcal{C}}$ of Hausdorff topologized semigroups.

A *topologized semigroup* is a topological space S endowed with an associative binary operation $S \times S \to S$, $(x, y) \mapsto xy$. If the binary operation is (separately) continuous, then S is called a *(semi) topological semigroup*. A topologized semigroup S is called *powertopological* if it is semitopological and for every $n \in \mathbb{N}$ the map $S \to S$, $x \mapsto x^n$, is continuous. A topologized semigroup S is called *right − topological* if for every $a \in S$ the right shift $S \to S$, $x \mapsto xa$, is continuous.

All topologized semigroups considered in this paper (except for those in Proposition 10 and Example 2) *are assumed to be Hausdorff*.

Topologized semigroups are objects of many categories which differ by morphisms. The most obvious category for morphisms has continuous homomorphisms between topologized semigroups. A bit wider category for morphisms has partial homomorphisms, i.e., homomorphisms defined on subsemigroups. The widest category for morphisms has semigroup relations. By a *semigroup relation* between semigroups X, Y we understand a subsemigroup $R \subset X \times Y$ of the product semigroup $X \times Y$.

Now we recall some standard operations on (semigroup) relations. For two (semigroup) relations $\Phi \subset X \times Y$ and $\Psi \subset Y \times Z$ their composition is the (semigroup) relation $\Psi \circ \Phi \subset X \times Z$ defined by $\Psi \circ \Phi = \{(x, z) \in X \times Z : \exists y \in Y \ (x, y) \in \Phi, \ (y, z) \in \Psi\}$. For a (semigroup) relation $R \subset X \times Y$ its inverse R^{-1} is the (semigroup) relation $R^{-1} = \{(y, x) : (x, y) \in R\} \subset Y \times X$.

For a relation $R \subset X \times Y$ and subsets $A \subset X$ the set $R(A) = \{y \in X : \exists a \in A \ (a, y) \in R\}$ is the *image* of A under the relation R. If $R \subset X \times Y$ is a semigroup relation between semigroups X, Y, then for any subsemigroup $A \subset X$ its image $R(A)$ is a subsemigroup of Y. For a relation $R \subset X \times Y$ the sets $R(X)$ and $R^{-1}(Y)$ are called the *range* and *domain* of R, respectively.

Semigroup relations between semigroups can be equivalently viewed as multimorphisms. By a *multimorphism* between semigroups X, Y we understand a multi-valued function $\Phi : X \multimap Y$ such that $\Phi(x) \cdot \Phi(y) \subset \Phi(xy)$ for any $x, y \in X$. Observe that a multi-valued function $\Phi : X \multimap Y$ between semigroups is a multimorphism if and only if its graph $\Gamma = \{(x, y) \in X \times Y : y \in \Phi(x)\}$ is a subsemigroup in $X \times Y$. Conversely, each subsemigroup $\Gamma \subset X \times Y$ determines a multimorphism

$\Phi : X \multimap Y, \Phi : x \mapsto \Phi(x) := \{(y \in Y : (x,y) \in \Gamma\}$. In the sequel we shall identify multimorphisms with their graphs.

A multimorphism $\Phi : X \multimap Y$ between semigroups X, Y is called a *partial homomorphism* if for each $x \in X$ the set $\Phi(x)$ contains at most one point. Each partial homomorphism $\Phi : X \multimap Y$ can be identified with the unique function $\varphi : \mathrm{dom}(\varphi) \to Y$ such that $\Phi(x) = \{\varphi(x)\}$ for each $x \in \mathrm{dom}(\varphi) := \Phi^{-1}(Y)$. This function φ is a homomorphism from the subsemigroup $\mathrm{dom}(\varphi)$ to the semigroup Y.

For a class $\mathcal{C}$ of Hausdorff topologized semigroups by c:$\mathcal{C}$ we denote the category whose objects are topologized semigroups in the class $\mathcal{C}$ and morphisms are closed semigroup relations between the topologized semigroups in the class $\mathcal{C}$. The category c:$\mathcal{C}$ contains the subcategories e:$\mathcal{C}$, i:$\mathcal{C}$, h:$\mathcal{C}$, and p:$\mathcal{C}$ whose objects are topologized semigroups in the class $\mathcal{C}$ and morphisms are isomorphic topological embeddings, injective continuous homomorphisms, continuous homomorphisms, and partial continuous homomorphisms with closed domain, respectively.

In this paper, we consider some concrete instances of the following general notion.

Definition 1. *Let $\vec{\mathcal{C}}$ be a category of topologized semigroups and their semigroup relations. A topologized semigroup X is called $\vec{\mathcal{C}}$-closed if for any morphism $\Phi \subset X \times Y$ of the category $\vec{\mathcal{C}}$ the range $\Phi(X)$ is closed in Y.*

In particular, for a class $\mathcal{C}$ of topologized semigroups, a topologized semigroup X is called

- e:$\mathcal{C}$-*closed* if for each isomorphic topological embedding $f : X \to Y \in \mathcal{C}$ the image $f(X)$ is closed in Y;
- i:$\mathcal{C}$-*closed* if for any injective continuous homomorphism $f : X \to Y \in \mathcal{C}$ the image $f(X)$ is closed in Y;
- h:$\mathcal{C}$-*closed* if for any continuous homomorphism $f : X \to Y \in \mathcal{C}$ the image $f(X)$ is closed in Y;
- p:$\mathcal{C}$-*closed* if for any continuous homomorphism $f : Z \to Y \in \mathcal{C}$ defined on a closed subsemigroup $Z \subset X$ the image $f(Z)$ is closed in Y;
- c:$\mathcal{C}$-*closed* if for any topologized semigroup $Y \in \mathcal{C}$ and any closed subsemigroup $\Phi \subset X \times Y$ the range $\Phi(X) := \{y \in Y : \exists x \in X \ (x,y) \in \Phi\}$ of Φ is closed in Y.

It is clear that for any class $\mathcal{C}$ of Hausdorff topologized semigroups and a topologized semigroup X we have the implications:

$$\text{c:}\mathcal{C}\text{-closed} \Rightarrow \text{p:}\mathcal{C}\text{-closed} \Rightarrow \text{h:}\mathcal{C}\text{-closed} \Rightarrow \text{i:}\mathcal{C}\text{-closed} \Rightarrow \text{e:}\mathcal{C}\text{-closed}.$$

In this paper, we are interested in characterizing topological groups which are e:$\mathcal{C}$-, i:$\mathcal{C}$-, h:$\mathcal{C}$-, p:$\mathcal{C}$- or c:$\mathcal{C}$-closed for the following classes of Hausdorff topologized semigroups:

- TS of all topological semigroups,
- pTS of all powertopological semigroups,
- sTS of all semitopological semigroups,
- rTS of all right-topological semigroups,

- TG of all topological groups,
- pTG of all paratopological groups,
- qTG of all quasitopological groups,
- sTG of all semitopological groups,
- rTG of all right-topological groups.

We recall that a *paratopological group* is a group G endowed with a topology making it a topological semigroup. So, the inversion operation is not necessarily continuous. A *quasitopological group* is a topologized group G such that for any $a, b \in G$ and $n \in \{1, -1\}$ the map $G \to G$, $x \mapsto ax^n b$, is continuous.

The inclusion relations between the classes of topologized semigroups are described in the following diagram (in which an arrow A $\to$ B between classes A, B indicates that A $\subset$ B).

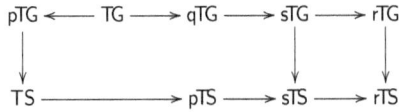

$$pTG \longleftarrow TG \longrightarrow qTG \longrightarrow sTG \longrightarrow rTG$$
$$\downarrow \qquad \qquad \qquad \qquad \downarrow \qquad \quad \downarrow$$
$$TS \longrightarrow pTS \longrightarrow sTS \longrightarrow rTS$$

In this paper we shall survey existing and new results related to the following general problem (consisting of $9 \times 5 = 45$ subproblems).

Problem 1. *Given a class* $\mathcal{C} \in \{TS, pTS, sTS, rTS, TG, qTG, pTG, sTG, rTG\}$ *and a class of morphisms* $f \in \{e, i, h, p, c\}$ *detect topological groups which are f:$\mathcal{C}$-closed.*

For the categories e:TG and e:qTG the answer to this problem is known and is a combined result of Raikov [1] who proved the equivalence (1) $\Leftrightarrow$ (3) and Bardyla, Gutik, Ravsky [2] who proved the equivalence (2) $\Leftrightarrow$ (3).

Theorem 1 (Raikov, Bardyla–Gutik–Ravsky). *For a topological group X the following conditions are equivalent:*

(1)   X is e:TG-closed;
(2)   X is e:qTG-closed;
(3)   X is complete.

A topological group X is *complete* if it is complete in its two-sided uniformity, i.e., the uniformity, generated by the entourages $\{(x, y) \in X \times X : y \in xU \cap Ux\}$ where U runs over neighborhoods of the unit in X.

On the other hand, Gutik ([3] 2.5) answered Problem 1 for the category e:sTS:

Theorem 2 (Gutik). *A topological group is compact if and only if it is e:sTS-closed.*

Theorems 1 and 2 and the trivial inclusions qTG $\supset$ TG $\subset$ pTG $\subset$ TS $\subset$ pTS $\subset$ sTS imply the following diagram of implications between various $\mathcal{C}$-closedness properties of a topological group:

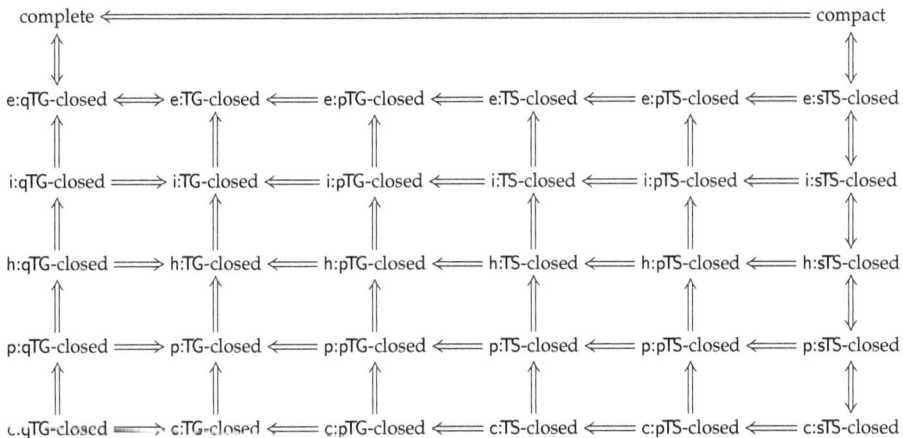

This diagram shows that various $\vec{C}$-closedness properties of topological groups fill and organize the "space" between compactness and completeness.

In fact, under different names, $\vec{C}$-closed topological (semi)groups have been already considered in mathematical literature. As we have already mentioned, e:TG-closed topological groups appeared in Raikov's characterization [1] of complete topological groups. The study of Lie groups which are i:TG-closed or h:TG-closed was initiated by Omori [4] in 1966 and continued by Goto [5] and currently by Bader and Gelander [6]. e:TS-Closed and h:TS-closed topological semigroups were introduced in 1969 by Stepp [7,8] who called them maximal and absolutely maximal semigroups, respectively. The study of h:TG-closed, p:TG-closed and c:TG-closed topological groups (called *h*-complete, hereditarily *h*-complete and *c*-compact topological groups, respectively) was initiated by Dikranjan and Tonolo [9] and continued by Dikranjan, Uspenskij [10], see the monograph of Lukàcs [11] and survey ([12] §4) of Dikranjan and Shakhmatov. The study of e:pTG-closed paratopological groups was initiated by Banakh and Ravsky [13,14], who called them *H*-closed paratopological groups. In [2,15–17] Hausdorff e:TS-closed (resp. h:TS-closed) topological semigroups are called (absolutely) *H*-closed. In [3] Gutik studied and characterized e:sTS-closed topological groups (calling them *H*-closed topological groups in the class of semitopological semigroups). The papers [18,19] are devoted to recognizing $\vec{C}$-closed topological semilattices for various categories $\vec{C}$ of topologized semigroups. In the paper [20] the author studied $\vec{C}$-closedness properties in Abelian topological groups and proved the following characterization (implying the famous Prodanov-Stoyanov Theorem on the precompactness of minimal Abelian topological groups, see [20]).

Theorem 3 (Banakh). *An Abelian topological group X is compact if and only if X is i:TG-closed.*

In Corollary 7 we shall complement this theorem proving that an Abelian topological group is compact if and only if it is e:sTG-closed.

The results of Banakh [20] and Ravsky [14] combined with Theorem 22 (proved in this paper) imply the following characterization of Abelian topological groups which are e:pTG-, e:TS- or e:pTS-closed.

Theorem 4 (Banakh, Ravsky). *For an Abelian topological group X the following conditions are equivalent:*

(1)   X is e:pTG-closed;
(2)   X is e:TS-closed;
(3)   X is e:pTS-closed;
(4)   X is complete and has compact exponent;
(5)   X is complete and for every injective continuous homomorphism $f : X \to Y$ to a topological group Y the group $\overline{f(X)}/f(X)$ is periodic.

A group X is called *periodic* if each element of X has finite order. Theorem 4(4) involves the (important) notion of a topological groups of compact exponent, which is defined as follows.

Definition 2. *A topological group X has (pre)compact exponent if for some $n \in \mathbb{N}$ the set $nX := \{x^n : x \in X\}$ has compact closure in X (resp. is totally bounded in X).*

Theorems 3 and 4 and Corollary 7 imply that for Abelian groups, the diagram describing the interplay between various $\vec{C}$-closedness properties collapses to the following form (containing only three different types of closedness: compactness, completeness, and completeness combined with compact exponent):

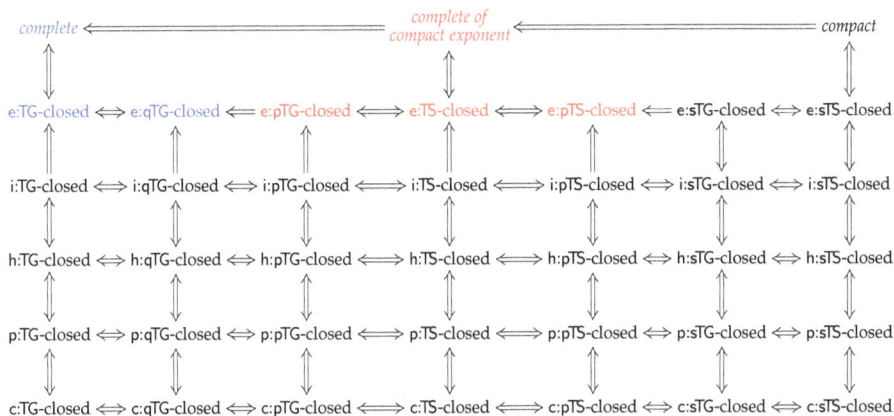

complete ⟸================⟹ complete of compact exponent ⟸================⟹ compact

e:TG-closed ⟺ e:qTG-closed ⟸ e:pTG-closed ⟺ e:TS-closed ⟺ e:pTS-closed ⟸ e:sTG-closed ⟺ e:sTS-closed

i:TG-closed ⟺ i:qTG-closed ⟺ i:pTG-closed ⟺ i:TS-closed ⟺ i:pTS-closed ⟺ i:sTG-closed ⟺ i:sTS-closed

h:TG-closed ⟺ h:qTG-closed ⟺ h:pTG-closed ⟹ h:TS-closed ⟺ h:pTS-closed ⟺ h:sTG-closed ⟺ h:sTS-closed

p:TG-closed ⟺ p:qTG-closed ⟺ p:pTG-closed ⟹ p:TS-closed ⟺ p:pTS-closed ⟺ p:sTG-closed ⟺ p:sTS-closed

c:TG-closed ⟺ c:qTG-closed ⟺ c:pTG-closed ⟺ c:TS-closed ⟺ c:pTS-closed ⟺ c:sTG-closed ⟺ c:sTS-closed

So, the problem remains to investigate the $\vec{\mathcal{C}}$-closedness properties for non-commutative topological groups. Now we survey the principal results (known and new) addressing this complex and difficult problem. We start with the following characterization of e:TS-closed topological groups, proved in Section 4.

Theorem 5. *A topological group X is e:TS-closed if and only if X is Weil-complete and for every continuous homomorphism $f : X \to Y$ into a Hausdorff topological semigroup Y the complement $\overline{f(X)} \setminus f(X)$ is not an ideal in the semigroup $\overline{f(X)}$.*

Using Theorems 3–5 we shall prove that various $\vec{\mathcal{C}}$-closedness properties have strong implications on the structure of subgroups related to commutativity, such as the subgroups of the topological derived series or the central series of a given topological group.

We recall that for a group G its *commutator* $[G, G]$ is the subgroup generated by the set $\{xyx^{-1}y^{-1} : x, y \in G\}$. The *topological derived series*

$$G = G^{[0]} \supset G^{[1]} \supset G^{[2]} \supset \cdots$$

of a topological group G consists of the subgroups defined by the recursive formula $G^{[n+1]} := \overline{[G^{[n]}, G^{[n]}]}$ for $n \in \omega$.

A topological group G is called *solvable* if $G^{[n]} = \{e\}$ for some $n \in \mathbb{N}$. The quotient group $X/X^{[1]}$ is called the *Abelianization* of a topological group X.

The *central series*

$$\{e\} = Z_0(G) \subset Z_1(G) \subset \cdots$$

of a (topological) group G consists of (closed) normal subgroups defined by the recursive formula

$$Z_{n+1}(G) := \{z \in G : \forall x \in G \ \ zxz^{-1}x^{-1} \in Z_n(G)\} \ \text{ for } \ n \in \omega.$$

A group G is called *nilpotent* if $G = Z_n(G)$ for some $n \in \omega$. The subgroup $Z_1(G)$ is called the *center* of the group G and is denoted by $Z(G)$.

The following theorem unifies Propositions 4 and Corollaries 6 and 9.

Theorem 6. *Let X be a topological group.*

(1)   *If X is e:pTG-closed, then the center $Z(X)$ has compact exponent.*

(2)   *If X is e:TS-closed, then for any closed normal subgroup $N \subset X$ the center $Z(X/N)$ of the quotient topological group X/N has precompact exponent.*

(3)   *If X is i:TG-closed or e:sTS-closed, then the center $Z(X)$ of X is compact.*

(4)   *If X is h:TG-closed, then for any closed normal subgroup $N \subset X$ the center $Z(X/N)$ is compact; in particular, the Abelianization $X/X^{[1]}$ of X is compact.*

Theorem 6(3), combined with an old result of Omori ([4] Corollary 1.3), implies the following characterization of i:TG-closed groups in the class of connected nilpotent Lie groups.

Theorem 7. *A connected nilpotent Lie group X is i:TG-closed if and only if X has compact center.*

Applying the statements (2) and (4) of Theorem 6 inductively, we obtain the following corollary describing the compactness properties of some characteristic subgroups of a $\vec{C}$-closed topological group (see Corollary 5, Proposition 7 and Theorem 35).

Corollary 1. *Let X be a topological group.*

(1)   *If X is e:TS-closed, then for every $n \in \omega$ the subgroup $Z_n(X)$ has compact exponent.*

(2)   *If X is h:TS-closed, then for every $n \in \omega$ the subgroup $Z_n(X)$ is compact.*

(3)   *If X is p:TG-closed, then for every $n \in \omega$ the quotient topological group $X/X^{[n]}$ is compact.*

The three items of Corollary 1 imply the following three characterizations. The first of them characterizes nilpotent complete group of compact exponent and is proved in Theorem 22.

Theorem 8. *For a nilpotent topological group X the following conditions are equivalent:*

(1)   *X is complete and has compact exponent;*

(2)   *X is e:TS-closed;*

(3)   *X is e:pTS-closed.*

In Example 2 we shall observe that the discrete topological group $\mathrm{Iso}(\mathbb{Z})$ of isometries of $\mathbb{Z}$ is e:TS-closed but does not have compact exponent. This shows that Theorem 8 does not generalize to solvable groups.

Theorem 9 (Dikranjan, Uspenskij). *For a nilpotent topological group X the following conditions are equivalent:*

(1)   *X is compact;*

(2)   *X is h:TG-closed.*

For Abelian topological groups Theorem 9 was independently proved by Zelenyuk and Protasov ([21]).

A topological group X is called *hypoabelian* if for each non-trivial closed subgroup X the commutator $[X, X]$ is not dense in X. It is easy to see that each solvable topological group is hypoabelian.

Theorem 10 (Dikranjan, Uspenskij). *For a solvable (more generally, hypoabelian) topological group X the following conditions are equivalent:*

(1)   *X is compact;*

(2)   *X is p:TG-closed;*

(3)   *any closed subgroup of X is h:TG-closed.*

The last two theorems were proved by Dikranjan and Uspenskij in ([10] 3.9 and 3.10) (in terms of the *h*-completeness, which is an alternative name for the h:TG-closedness).

The Weyl-Heisenberg group $H(w_0)$ (which is a non-compact i:TG-closed nilpotent Lie group) shows that h:TG-closedness in Theorem 9 cannot be weakened to the i:TG-closedness (see Example 1 for more details).

On the other hand, the solvable Lie group $\mathrm{Iso}_+(\mathbb{C})$ of orientation preserving isometries of the complex plane is h:TS-closed and not compact, which shows that the p:TG-closedness in Theorem 10(2) cannot be replaced by the h:TG-closedness of *X*. This example (analyzed in details in Section 10) answers Question 3.13 in [10] and Question 36 in [12].

Nonetheless, the p:TG-closedness of the solvable group *X* in Theorem 10(2) can be replaced by the h:TG-closedness of *X* under the condition that the group *X* is balanced and MAP-solvable.

A topological group *X* is called *balanced* if for any neighborhood $U \subset X$ of the unit there exists a neighborhood $V \subset X$ of the unit such that $xV \subset Ux$ for all $x \in X$. A topological group *X* is balanced if and only if the left and right uniformities on *X* coincide.

A topological group *X* is called *maximally almost periodic* (briefly *MAP*) if it admits a continuous injective homomorphism $h : X \to K$ into a compact topological group *K*. By Theorem 37, for any productive class $\mathcal{C} \supset$ TG of topologized semigroups, *the i:$\mathcal{C}$-closedness and h:$\mathcal{C}$-closedness are equivalent for MAP topological groups.*

A topological group *X* is defined to be *MAP-solvable* if there exists an increasing sequence $\{e\} = X_0 \subset X_1 \subset \cdots \subset X_m = X$ of closed normal subgroups in *X* such that for every $n < m$ the quotient group X_{n+1}/X_n is Abelian and MAP. Since locally compact Abelian groups are MAP, each solvable locally compact topological group is MAP-solvable.

The following theorem (proven in Section 9) nicely complements Theorem 10 of Dikranjan and Uspenskij. Example 3 of non-compact solvable h:TS-closed Lie group $\mathrm{Iso}_+(\mathbb{C})$ shows that the "balanced" requirement cannot be removed from the conditions (2), (3).

Theorem 11. *For a solvable topological group X the following conditions are equivalent:*

(1)   *X is compact;*

(2)   *X is balanced, locally compact, and h:TG-closed;*

(3)   *X is balanced, MAP-solvable and h:TG-closed.*

It is interesting that the proof of this theorem exploits a good piece of the descriptive set theory (that dealing with *K*-analytic spaces). Also methods of descriptive set theory are used for establishing the interplay between i:TG-closed and minimal topological groups.

We recall that a topological group *X* is *minimal* if each continuous bijective homomorphism $h : X \to Y$ onto a topological group *Y* is open (equivalently, is a topological isomorphism). By the fundamental theorem of Prodanov and Stoyanov [22], each minimal topological Abelian group is precompact, i.e., has compact Raikov completion. Groups that are minimal in the discrete topology are called *non-topologizable*. For more information on minimal topological groups we refer the reader to the monographs [11,23] and the surveys [24,25].

The definition of minimality implies that *a minimal topological group is i:TG-closed if and only if it is e:TG-closed if and only if it is complete.* In particular, each minimal complete topological group is i:TG-closed. By Theorem 3, the converse implication holds for Abelian topological groups. It also holds for ω-narrow topological groups of countable pseudocharacter.

Theorem 12. *An ω-narrow topological group X of countable pseudocharacter is i:TG-closed if and only if X is complete and minimal.*

A subset $B \subset X$ of a topological group X is called *ω-narrow* if for any neighborhood $U \subset X$ of the unit there exists a countable set $C \subset X$ such that $B \subset CU \cap UC$. ω-Narrow topological groups were introduced by Guran [26] (as $\aleph_0$-bounded groups) and play important role in the theory of topological groups [27]. Theorem 12 will be proved in Section 5 (see Theorem 31). This theorem suggests the following open problem.

Problem 2. *Is each* i:TG-*closed topological group minimal?*

Observe that a complete MAP topological group is minimal if and only if it is compact. So, for MAP topological groups Problem 2 is equivalent to another intriguing open problem.

Problem 3. *Is each* i:TG-*closed MAP topological group compact?*

For ω-narrow topological groups an affirmative answer to this problem follows from Theorem 37 and the characterization of h:TG-closedness in term of total completeness and total minimality, see Theorem 13.

Following [11], we define a topological group G to be *totally complete* (resp. *totally minimal*) if for any closed normal subgroup $H \subset G$ the quotient topological group G/H is complete (resp. minimal). Totally minimal topological groups were introduced by Dikranjan and Prodanov in [28]. By ([11] 3.45), each totally complete totally minimal topological group is absolutely TG-closed.

Theorem 13. *An ω-narrow topological group is* h:TG-*closed if and only if it is totally complete and totally minimal.*

Theorem 13 will be proved in Section 6 (see Theorem 33). This theorem complements a characterization of h:TG-closed topological groups in terms of special filters, due to Dikranjan and Uspenskij [10] (see also [11] 4.24). Using their characterization of h:TG-closedness, Dikranjan and Uspenskij [10] proved another characterization.

Theorem 14 (Dikranjan, Uspenskij). *A balanced topological group is* h:TG-*closed if and only if it is* c:TG-*closed.*

The compactness of ω-narrow i:TG-closed MAP topological groups can be also derived from the compactness of the *ω-conjucenter* $Z^\omega(X)$ defined for any topological group X as the set of all points $z \in X$ whose conjugacy class $C_X(z) := \{xzx^{-1} : x \in X\}$ is ω-narrow in X.

A topological group X is defined to be *ω-balanced* if for any neighborhood $U \subset X$ of the unit there exists a countable family $\mathcal{V}$ of neighborhoods of the unit such that for any $x \in X$ there exists $V \in \mathcal{V}$ such that $xVx^{-1} \subset U$. It is known (and easy to see) that each ω-narrow topological group is ω-balanced. By Katz Theorem [27], a topological group is ω-balanced if and only if it embeds into a Tychonoff product of first-countable topological groups. The following theorem can be considered as a step towards the solution of Problem 3.

Theorem 15. *If an ω-balanced MAP topological group X is* i:TG-*closed, then its ω-conjucenter $Z^\omega(X)$ is compact.*

A topological group G is called *hypercentral* if for each closed normal subgroup $H \subsetneq G$, the quotient group G/H has non-trivial center $Z(G/H)$. It is easy to see that each nilpotent topological group is hypercentral. Theorem 15 implies the following characterization (see Corollary 17).

Corollary 2. *A hypercentral topological group X is compact if and only if X is ω-balanced, MAP, and* i:TG-*closed.*

Remark 1. *Known examples of non-topologizable groups (due to Klyachko, Olshanskii, and Osin [29]) show that the compactness does not follow from the pTS- or c:TG-closedness even for 2-generated discrete topological groups (see Example 4).*

The following diagram describes the implications between various completeness and closedness properties of a topological group. By simple arrows we indicate the implications that hold under some additional assumptions (written in italic near the arrow).

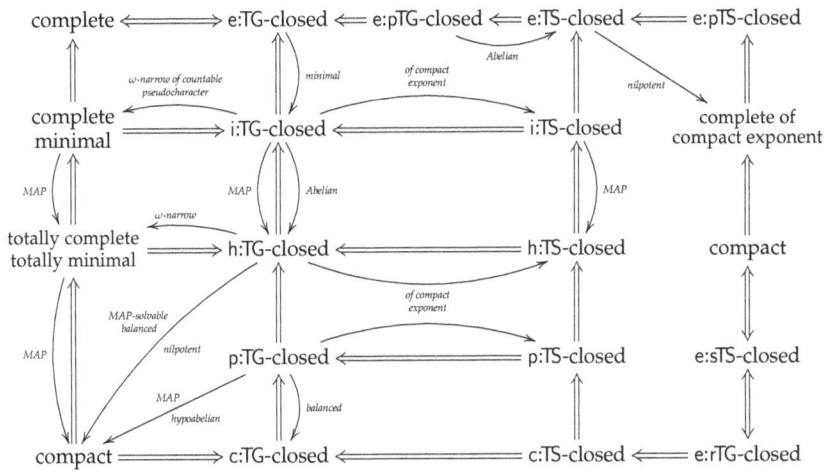

The curved horizontal implications, holding under the assumption of compact exponent, are proved in Theorems 24 and 32.

2. Completeness of Topological Groups Versus $\mathcal{C}$-closedness

To discuss the completeness properties of topological groups, we need to recall some known information related to uniformities on topological groups (see [27,30] for more details). We refer the reader to ([31] Ch.8) for basic information on uniform spaces. Here we recall that a uniform space $(X, \mathcal{U})$ is *complete* if each Cauchy filter $\mathcal{F}$ on X converges to some point $x \in X$. A *filter* on a set X is a non-empty family of non-empty subsets of X, which is closed under finite intersections and taking supersets. A subfamily $\mathcal{B} \subset \mathcal{F}$ is called a *base* of a filter $\mathcal{F}$ if each set $F \in \mathcal{F}$ contains some set $B \in \mathcal{B}$.

A filter $\mathcal{F}$ on a uniform space $(X, \mathcal{U})$ is *Cauchy* if for each entourage $U \in \mathcal{U}$ there is a set $F \in \mathcal{F}$ such that $F \times F \subset U$. A filter on a topological space X *converges* to a point $x \in X$ if each neighborhood of x in X belongs to the filter. A uniform space $(X, \mathcal{U})$ is compact if and only if the space is complete and *totally bounded* in the sense that for every entourage $U \in \mathcal{U}$ there exists a finite subset $F \subset X$ such that $X = \bigcup_{x \in F} B(x, U)$ where $B(x, U) := \{y \in X : (x, y) \in U\}$.

Each topological group (X, τ) with unit e carries four natural uniformities:

- the *left uniformity* $\mathcal{U}_l$ generated by the base $\mathcal{B}_l = \{\{(x, y) \in X \times X : y \in xU\} : e \in U \in \tau\}$;
- the *right uniformity* $\mathcal{U}_r$ generated by the base $\mathcal{B}_r = \{\{(x, y) \in X \times X : y \in Ux\} : e \in U \in \tau\}$;
- the *two-sided uniformity* $\mathcal{U}_\vee$ generated by the base $\mathcal{B}_\vee = \{\{(x, y) \in X \times X : y \in Ux \cap xU\} : e \in U \in \tau\}$;
- the *Roelcke uniformity* $\mathcal{U}_\wedge$ generated by the base $\mathcal{B}_\wedge = \{\{(x, y) \in X \times X : y \in UxU\} : e \in U \in \tau\}$.

It is well-known (and easy to see) that a topological group X is complete in its left uniformity if and only if it is complete in its right uniformity. Such topological groups are called *Weil-complete*. A topological group is *complete* if it is complete in its two-sided uniformity. Since each Cauchy filter in the two-sided uniformity is Cauchy in the left and right uniformities, each Weil-complete topological group is complete. For an Abelian (more generally, balanced) topological group X all four uniformities $\mathcal{U}_l, \mathcal{U}_r, \mathcal{U}_\vee, \mathcal{U}_\wedge$ coincide, which implies that X is Weil-complete if and only if it is complete.

An example of a complete topological group, which is not Weil-complete is the Polish group $\mathrm{Sym}(\omega)$ of all bijections of the discrete countable space ω (endowed with the topology of pointwise convergence, inherited from the Tychonoff product ω^ω).

The completion of a topological group X by its two-sided uniformity is called the *Raikov-completion* of X. It is well-known that the Raikov-completion of a topological group has a natural structure of a topological group, which contains X as a dense subgroup. On the other hand, the completion of a topological group X by its left (or right) uniformity carries a natural structure of a topological semigroup, called the (left or right) *Weil-completion* of the topological group, see ([32] 8.45). For example, the left Weil-completion of the Polish group $\mathrm{Sym}(\omega)$ is the semigroup of all injective functions from ω to ω.

So, if a topological group X is not complete, then X admits a non-closed embedding into its Raikov-completion, which implies that it is not e:$\mathcal{C}$-closed for any class $\mathcal{C}$ of topologized semigroups, containing all complete topological groups. If X is not Weil-complete, then X admits a non-closed embedding into its (left or right) Weil-completion, which implies that it is not e:$\mathcal{C}$-closed for any class $\mathcal{C}$ of topologized semigroups, containing all Tychonoff topological semigroups. Let us write these facts for future references.

Theorem 16. *Assume that a class $\mathcal{C}$ of topologized semigroups contains all Raikov-completions (and Weil-completions) of topological groups. Each e:$\mathcal{C}$-closed topological group is (Weil-)complete. In particular, each e:TG-closed topological group is complete and each e:TS-closed topological group is Weil-complete.*

We recall that a non-empty subset I of a semigroup S is called an *ideal* in S if $IS \cup SI \subset I$.

Theorem 17. *Assume that a topological group X admits a non-closed topological isomorphic embedding $f : X \to Y$ into a Hausdorff semitopological semigroup Y.*

(1)   *If X is Weil-complete, then $\overline{f(X)} \setminus f(X)$ is an ideal of the semigroup $\overline{f(X)}$.*
(2)   *If X is complete, then $\{y^n : y \in \overline{f(X)} \setminus f(X), \ n \in \mathbb{N}\} \subset \overline{f(X)} \setminus f(X)$.*

Proof. To simplify notation, it will be convenient to identify X with its image $f(X)$ in Y. Replacing Y by the closure $\bar{X}$ of X, we can assume that the group X is dense in the semigroup Y.

1. First, we assume that X is Weil-complete. Given any $x \in X$ and $y \in Y \setminus X$, we should prove that $xy, yx \in Y \setminus X$.

To derive a contradiction, assume that $xy \in X$. On the topological group X, consider the filter $\mathcal{F}$ generated by the base consisting of the intersections $X \cap O_y$ of X with neighborhoods O_y of y in Y. The Hausdorff property of Y ensures that this filter does not converge in the Weil-complete group X and thus is not Cauchy in the left uniformity of X. This yields an open neighborhood V_e of the unit of the group X such that $F \not\subset zV_e$ for any set $F \in \mathcal{F}$ and any $z \in X$. Since X carries the subspace topology, the space Y contains an open set $U_{xy} \subset Y$ such that $U_{xy} \cap X = xyV_e$.

The separate continuity of the binary operation on Y yields an open neighborhood $U_x \subset Y$ of the point x in Y such that $U_x y \subset U_{xy}$. Choose any point $z \in U_x$ and find a neighborhood U_y of the point y in Y such that $zU_y \subset U_{xy}$. Now consider the set $F = X \cap U_y \in \mathcal{F}$ and observe that $zF \subset X \cap U_{xy} = xyV_e$ and hence $F \subset z^{-1}xyV_e$, which contradicts the choice of V_e. This contradiction shows that $xy \in Y \setminus X$.

By analogy we can prove that $yx \in Y \setminus X$.

2. Next, assume that X is complete. In this case we should prove that $y^n \notin X$ for any $y \in Y \setminus X$ and $n \in \mathbb{N}$. To derive a contradiction, assume that $y^n \in X$ for some $y \in Y \setminus X$ and $n \in \mathbb{N}$. On the group X consider the filter $\mathcal{F}$ generated by the base $X \cap O_y$ where O_y runs over neighborhoods of y in Y. The filter $\mathcal{F}$ converges to the point $y \notin X$ and hence is divergent in X (by the Hausdorff property of Y).

Since X is complete, the divergent filter $\mathcal{F}$ is not Cauchy in its two-sided uniformity. This allows us to find an open neighborhood $V_e \subset X$ of the unit such that $F \not\subset xV_e \cap V_e z$ for any points $x, z \in X$. Choose an open set $W \subset Y$ such that $W \cap X = y^n V_e \cap V_e y^n$.

By finite induction, we shall construct a sequence $(x_i)_{i=0}^{n-1}$ of points of the group X such that $x_i y^{n-i} \in W$ for all $i \in \{1, \dots, n-1\}$. To start the inductive construction, let $x_0 = e$ be the unit of the group X. Assume that for some positive $i \le n-1$ the point $x_{i-1} \in X$ with $x_{i-1} y^{n+1-i} \in W$ has been constructed. By the separate continuity of the semigroup operation in Y, the point y has a neighborhood $V_y \subset Y$ such that $x_{i-1} V_y y^{n-i} \in W$. Choose any point $v \in X \cap V_y$, put $x_i = x_{i-1} v \in X \cap V_y$ and observe that $x_i y^{n-i} \in W$, which completes the inductive step.

After completing the inductive construction, we obtain a point $x = x_{n-1} \in X$ such that $xy \in W$. By analogy we can construct a point $z \in X$ such that $yz \in W$. The separate continuity of the binary operation in Y yields a neighborhood $V_y \subset Y$ of y such that $(xV_y) \cup (V_y z) \subset W$. Then the set $F = X \cap V_y \in \mathcal{F}$ has the property: $(xF) \cup (Fz) \subset X \cap W = y^n V_e \cap V_e y^n$ which implies that $F \subset (x^{-1} y^n V_e) \cap (V_e y^n z^{-1})$. However, this contradicts the choice of the neighborhood V_e. $\square$

Now we describe a construction of the ideal union of topologized semigroups, which allows us to construct non-closed embeddings of topologized semigroups.

Let $h : X \to Y$ be a continuous homomorphism between topologized semigroups X, Y such that $Y \setminus h(X)$ is an ideal in Y and $X \cap (Y \setminus h(X)) = \emptyset$. Consider the set $U_h(X, Y) := X \cup (Y \setminus h(X))$ endowed with the semigroup operation defined by

$$
xy = \begin{cases}
x * y & \text{if } x, y \in X; \\
h(x) \cdot y & \text{if } x \in X \text{ and } y \in Y \setminus h(X); \\
x \cdot h(y) & \text{if } x \in Y \setminus h(X) \text{ and } y \in X; \\
x \cdot y & \text{if } x, y \in Y \setminus h(X).
\end{cases}
$$

Here by $*$ and $\cdot$ we denote the binary operations of the semigroups X and Y, respectively. The set $U_h(X, Y)$ is endowed with the topology consisting of the sets $W \subset U_h(X, Y)$ such that

- for any $x \in W \cap X$, some neighborhood $U_x \subset X$ of x is contained in W;
- for any $y \in W \cap (Y \setminus h(X))$ there exists an open neighborhood $U \subset Y$ of y such that $h^{-1}(U) \cup (U \setminus h(X)) \subset W$.

This topology turns $U_h(X, Y)$ into a topologized semigroup, which be called the *ideal union of the semigroups X and Y along the homomorphism h*.

The following theorem can be derived from the definition of the ideal union.

Theorem 18. *Let $h : X \to Y$ be a continuous homomorphism between topologized semigroups such that $Y \setminus h(X)$ is an ideal in Y and $(Y \setminus h(X)) \cap X = \emptyset$. The topologized semigroup $U_h(X, Y)$ has the following properties:*

(1)   *X is an open subsemigroup of $U_h(X, Y)$;*
(2)   *X is closed in $U_h(X, Y)$ if and only if $h(X)$ is closed in Y;*
(3)   *If X and Y are (semi)topological semigroups, then so is the topologized semigroup $U_h(X, Y)$;*
(4)   *If the spaces X, Y are Hausdorff (or regular or Tychonoff), then so is the space $U_h(X, Y)$.*

We shall say that a class $\mathcal{C}$ of topologized semigroups is stable under taking

- *topological isomorphisms* if for any topological isomorphism $h : X \to Y$ between topologized semigroups X, Y the inclusion $X \in \mathcal{C}$ implies $Y \in \mathcal{C}$;
- *closures* if for any topologized semigroup $Y \in \mathcal{C}$ and a subgroup $X \subset Y$ the closure $\bar{X}$ of X in Y is a topologized semigroup that belongs to the class $\mathcal{C}$;

- *ideal unions* if for any continuous homomorphism $h : X \to Y$ between semigroups $X, Y \in \mathcal{C}$ with $Y \setminus h(X)$ being an ideal in Y, disjoint with X, the topologized semigroup $U_h(X, Y)$ belongs to the class $\mathcal{C}$.

Theorems 17 and 18 imply the following characterization.

Theorem 19. *Assume that a class $\mathcal{C}$ of Hausdorff semitopological semigroups is stable under topological isomorphisms, closures and ideal unions. A Weil-complete topological group $X \in \mathcal{C}$ is e:$\mathcal{C}$-closed if and only if for every continuous homomorphism $f : X \to Y$ into a topologized semigroup $Y \in \mathcal{C}$ the set $\overline{f(X)} \setminus f(X)$ is not an ideal in $\overline{f(X)}$.*

Proof. To prove the "only if" part, assume that there exits a continuous homomorphism $f : X \to Y$ into a topologized semigroup $Y \in \mathcal{C}$ such that $f(X)$ is dense in Y and $\overline{f(X)} \setminus f(X)$ is an ideal of $\overline{f(X)}$. In particular, $\overline{f(X)} \setminus f(X) \neq \emptyset$, which means that $f(X)$ is not closed in Y. Taking into account that the class $\mathcal{C}$ is stable under closures, we conclude that $\overline{f(X)}$ is a topologized semigroup in the class $\mathcal{C}$. So, we can replace Y by $\overline{f(X)}$ and assume that the subgroup $f(X)$ is dense in Y. Replacing Y by its isomorphic copy, we can assume that $X \cap Y = \emptyset$. In this case we can consider the ideal sum $U_f(X, Y)$ and conclude that it belongs to the class $\mathcal{C}$ (since $\mathcal{C}$ is stable under ideal unions). By Theorem 18(2), the topological group X is not closed in $U_f(X, Y)$, which means that X is not e:$\mathcal{C}$-closed.

To prove the "if" part, assume that the Weil-complete topological group X is not e:$\mathcal{C}$-closed. Then X admits a non-closed topological isomorphic embedding $f : X \to Y$ into a topologized semigroup $Y \in \mathcal{C}$. By Theorem 17(1), the complement $\overline{f(X)} \setminus f(X)$ is an ideal in $\overline{f(X)}$. □

3. Topological Groups of (Pre)compact Exponent

In this section, we study topological groups of compact exponent. We shall say that a topological group X has *compact exponent* (resp. *finite exponent*) if there exists a number $n \in \mathbb{N}$ such that the set $nX := \{x^n : x \in X\}$ is contained is a compact (resp. finite) subset of X. A complete topological group has compact exponent if and only if it has *precompact exponent* in the sense that for some $n \in \mathbb{N}$ the set nX is precompact. A subset A of a topological group X is called *precompact* if for any neighborhood $U \subset X$ of the unit there exists a finite subset $F \subset X$ such that $A \subset FU \cap UF$.

Lemma 1. *A topological group X is precompact if and only if for any neighborhood $U \subset X$ of the unit there exists a finite subset $F \subset X$ such that $X = FUF$.*

Proof. The "only if" part is trivial. To prove the "if" part, assume that for any neighborhood $U \subset X$ of the unit there exists a finite subset $F \subset X$ such that $A \subset FUF$. Given a neighborhood $W = W^{-1} \subset X$ of the unit, we need to find a finite subset $E \subset X$ such that $X = EW = WE$. Choose a neighborhood $U \subset X$ of the unit such that $UU^{-1} \subset W$. By our assumption, there exists a finite set $F \subset X$ such that $X = FUF$. By ([33] 12.6), for some $x, y \in F$ there exists a finite subset $B \subset G$ such that $G = B(xUy)(xUy)^{-1}$. Then $G = BxUU^{-1}x^{-1}$ and hence $G = EW = WE^{-1}$ for $E = Bx$. □

The following proposition shows that our definition of a group of finite exponent is equivalent to the standard one.

Proposition 1. *A group X has finite exponent if and only if there exists $n \in \mathbb{N}$ such that for every $x \in X$ the power x^n coincides with the unit of the group X.*

Proof. The "if" part is trivial. To prove the "only if" part, assume that X has finite exponent and find $n \in \mathbb{N}$ such that the set $F = \{x^n : x \in X\}$ is finite.

It follows that for every $x \in F$ the powers x^{kn}, $k \in \mathbb{N}$, belong to the set F. So, by the Pigeonhole Principle, $x^{in} = x^{jn}$ for some numbers $i < j$. Consequently, for the number $m_x = j - i$ the power x^{nm_x}

is the unit e of the group X. Then for the number $m = \prod_{x \in F} m_x$ we have $\{x^{n^2 m} : x \in X\} \subset \{x^{nm} : x \in F\} = \{e\}$. $\square$

This characterization implies that being of finite exponent is a 3-space property.

Corollary 3. *Let H be a normal subgroup of a group G. The group G has finite exponent if and only if H and G/H have finite exponent.*

A similar 3-space property holds also for topological groups of compact exponent. A subgroup H of a group G is called *central* if H is contained in the *center* $Z(G) = \{x \in G : \forall g \in G \ xg = gx\}$ of the group G.

Proposition 2. *Let Z be a closed central subgroup of a topological group X. The topological group X has precompact exponent if and only if the topological groups Z and $Y = X/Z$ have precompact exponent.*

Proof. If the topological group X has precompact exponent, then for some $n \in \mathbb{N}$ the set $nX = \{x^n : n \in X\}$ is precompact in X. It follows that the intersection $(nX) \cap Z \supset nZ$ is precompact in Z and the image $q(nX) = nY$ of nX under the quotient homomorphism $q : X \to X/Z = Y$ is precompact in the quotient topological group Y.

The proof of the "if" part is more complicated. Assume that the topological groups Z and $Y := X/Z$ have precompact exponent. Then there exist natural numbers n and m such that the sets $A := nZ$ and $B := mY$ are precompact. The set B contains the unit of the group Y and hence $B^k \subset B^n$ for all positive $k \le n$.

We claim that the set $nmX := \{x^{nm} : x \in X\}$ is precompact in X. Given any open neighborhood $U = U^{-1} \subset G$ of the unit, we should find a finite subset $F \subset G$ such that the set $nmX \subset FU \cap UF$. Since the set nmX is symmetric, it suffices to find $F \subset X$ such that $nmX \subset FU$. Using the continuity of the group operations, choose a neighborhood $V \subset X$ of the unit such that $V = V^{-1}$ and $V^{3n+1} \subset U$. Let $q : X \to Y$ be the quotient homomorphism.

Claim 1. *For the precompact set B^n the intersection $W = \bigcap_{x \in q^{-1}(B^n)} x V^3 x^{-1}$ is a neighborhood of the unit in X.*

Proof. By the precompactness of B^n and the openness of the quotient homomorphism $q : X \to X/Z$, there exists a finite set $F \subset X$ such that $B^n \subset q(FV)$. We claim that the neighborhood $W' = \bigcap_{y \in F} y V y^{-1}$ is contained in $x V^3 x^{-1}$ for any $x \in q^{-1}(B^n)$. Indeed, for any $w \in W'$ and $x \in q^{-1}(B^n)$, we can find $y \in F$ such that $q(x) \in q(yV)$ and hence $x = yvz$ for some $v \in V$ and $z \in Z$. Then $w \in y V y^{-1} \subset xz^{-1}v^{-1}Vvzx^{-1} = xv^{-1}Vvx^{-1} \subset x V^3 x^{-1}$ (here we use that z belongs to the center of the group X). $\square$

By the precompactness of the sets A, B, there exist a finite set $A' \subset A \subset Z$ such that $A \subset A'V$ and a finite set $B' \subset q^{-1}(B)$ such that $B \subset q(B') \cdot q(W) = q(B'W)$. We claim that the finite set $F = \{b^n a : b \in B', \ a \in A'\}$ has the desired property: $x^{mn} \in FU$ for any $x \in X$.

The choice of m ensures that $x^m \in q^{-1}(B) \subset q^{-1}(q(B'W)) = B'WZ$. So, we can find elements $b \in B', w \in W$ and $z \in Z$ such that $x^m = bwz$ and hence $x^{mn} = (bwz)^n = (bw)^n z^n \in (bw)^n A \subset (bw)^n A'V$ (we recall that the element $z \in Z$ belongs to the center of X).

Observe that $(bw)^n = \left(\prod_{i=1}^n b^i w b^{-i}\right) b^n$. For every $i \le n$ the element b^i belongs to $q^{-1}(B^n)$ and by Claim 1, $b^i w b^{-i} \in b^n V^3 b^{-n}$. So,

$$x^{mn} \in (bw)^n A'V = \left(\prod_{i=1}^n b^i w b^{-i}\right) b^n A'V \subset (b^n V^3 b^{-n})^n b^n A'V = b^n V^{3n} A'V = b^n A' V^{3n+1} \subset FU.$$

$\square$

For complete topological groups, the precompactness of exponent is recognizable by countable subgroups.

Proposition 3. *A complete topological group X has precompact exponent if and only if each countable subgroup of X has precompact exponent.*

This proposition can be easily derived from the following (probably known) lemma.

Lemma 2. *A subset A of a topological group X is precompact if and only if for each countable subgroup $H \subset X$ the intersection $A \cap H$ is precompact in the topological group H.*

Proof. The "only if" part is trivial. To prove the "only if" part, assume that A is not precompact. Then $A \cup A^{-1}$ is not precompact and hence there exists a neighborhood $U = U^{-1} \subset X$ of the unit such that $A \cup A^{-1} \neq FU$ for any finite subset $F \subset X$. By Zorn's Lemma, there exists a maximal subset $E \subset A \cup A^{-1}$ which is U-separated in the sense that $x \notin yU$ for any distinct points $x, y \in E$. The maximality of E guarantees that for any $x \in A \cup A^{-1}$ there exists $y \in E$ such that $x \in yU$ or $y \in xU$ (and hence $x \in yU^{-1} = yU$). Consequently, $A \cup A^{-1} = EU$. The choice of U ensures that the set E is infinite. Then we can choose any infinite countable set $E_0 \subset E$ and consider the countable subgroup H generated by E_0. It follows that the intersection $H \cap (A \cup A^{-1})$ containes the infinite U-separated set E_0 and hence is not precompact in H. □

4. On e:$\mathcal{C}$-closed Topological Groups

In this section, we collect some results on e:$\mathcal{C}$-closed topological groups for various classes $\mathcal{C}$.

First, observe that Theorems 16, 17 and 19 imply the following theorem (announced as Theorem 5 in the introduction).

Theorem 20. *A topological group X is e:TS-closed if and only if X is Weil-complete and for every continuous homomorphism $f : X \to Y$ into a Hausdorff topological semigroup Y the complement $\overline{f(X)} \setminus f(X)$ is not an ideal in the semigroup $\overline{f(X)}$.*

We recall that a topologized semigroup X is defined to be a *powertopological semigroup* if it is semitopological and for every $n \in \mathbb{N}$ the power map $X \to X, x \mapsto x^n$, is continuous. By pTS we denote the class of Hausdorff powertopological semigroups.

Theorem 21. *Each complete topological group X of compact exponent is e:pTS-closed.*

Proof. Fix a number $n \in \mathbb{N}$ and a compact set $K \subset X$ such that $\{x^n : x \in X\} \subset K$. To show that X is e:pTS-closed, assume that X is a subgroup of some Hausdorff powertopological semigroup Y. The Hausdorff property of Y ensures that the compact set K is closed in Y. Then the continuity of the power map $p : Y \to Y, p : y \mapsto y^n$, implies that the set

$$\{y \in Y : y^n \in K\}$$

containing X is closed in Y and hence contains $\bar{X}$. If X is not closed in Y, then we can find a point $y \in \bar{X} \setminus X$ and conclude that $y^n \in K \subset X$. However, this contradicts Theorem 17(2). □

Corollary 4. *For a topological group X of precompact exponent the following conditions are equivalent:*

(1)   *X is complete;*
(2)   *X is e:TG-closed;*
(3)   *X is e:TS-closed;*
(4)   *X is e:pTS-closed.*

Proof. The implications (4) ⇒ (3) ⇒ (2) follow from the inclusions pTS ⊃ TS ⊃ TG, (2) ⇒ (1) and (1) ⇒ (4) follow from Theorems 16 and 21, respectively. □

Proposition 4. *If a topological group X is e:TS-closed, then for any closed normal subgroup N ⊂ X the center Z(X/N) of the quotient group X/N has precompact exponent.*

Proof. Let $G = X/N$ be the quotient topological group, $q : X \to G$ be the quotient homomorphism and $Z = \{z \in Z : \forall g \in G \ zg = gz\}$ be the center of the group G. Assuming that Z does not have precompact exponent, we conclude that the completion $\bar{Z}$ of Z does not have compact exponent. Applying Theorem 20, we obtain a continuous injective homomorphism $f : \bar{Z} \to Y$ to a topological group Y such that the closure $\overline{f(\bar{Z})} = \overline{f(Z)}$ of $f(\bar{Z})$ in Y contains an element y such that $y^n \notin f(\bar{Z})$ for all $n \in \mathbb{N}$.

Observe that the family $\tau_Z = \{Z \cap f^{-1}(U) : U \text{ is open in } Y\}$ is a Hausdorff topology on Z turning it into a topological group, which is topologically isomorphic to the topological group $f(Z)$. Then the completion $\bar{Z}$ of the topological group (Z, τ_Z) contains an element $z \in \bar{Z}$ such that $z^n \notin Z$ for all $n \in \mathbb{N}$.

Let $\mathcal{T}_G$ be the topology of the topological group G. Taking into account that the subgroup Z is central in G, we can show that the family $\tau_e = \{U \cdot V : e \in U \in \mathcal{T}_G, \ e \in V \in \tau_Z\}$ satisfies the Pontryagin Axioms ([27] 1.3.12) and hence is a neighborhood base at the unit of some Hausdorff group topology τ on G. The definition of this topology implies that the subgroup Z remains closed in the topology τ and the subspace topology $\{U \cap Z : U \in \tau\}$ on Z coincides with the topology τ_Z. Then the completion $\bar{Z}$ of the topological group (Z, τ_Z) is contained in the completion $\bar{G}$ of the topological group (G, τ) and hence $z \in \bar{Z} \subset \bar{G}$. Now consider the subsemigroup S of $\bar{G}$, generated by the set $G \cup \{z\}$. Observe that $\{z^n\}_{n \in \omega} \subset \bar{Z} \setminus Z = \bar{Z} \setminus G$. Since the group Z is central in G, the element z commutes with all elements of G. This implies that $S = \{gz^n : g \in G, \ n \in \omega\}$ and hence $S \setminus G = \{gz^n : g \in G, \ n \in \mathbb{N}\}$ is an ideal in G. Let $i : G \to S$ be the identity homomorphism. Then for the homomorphism $h = i \circ q : X \to S$ the complement $\overline{h(X)} \setminus h(X) = S \setminus G$ is an ideal in S. By Theorem 20, the topological group X is not e:TS-closed. This is a desired contradiction showing that the topological group $Z(G) = Z(X/N)$ has precompact exponent. □

We recall that for a topological group X its central series $\{e\} = Z_0(X) \subset Z_1(X) \subset \cdots$ consists of the subgroups defined recursively as $Z_{n+1}(X) = \{z \in X : \forall x \in X \ zxz^{-1}x^{-1} \in Z_n(X)\}$ for $n \in \omega$.

Corollary 5. *If a topological group X is e:TS-closed, then for every $n \in \omega$ the subgroup $Z_n(X)$ has compact exponent.*

Proof. First observe that the topological group X is complete, being e:TS-closed. Then its closed subgroups $Z_n(X)$, $n \in \omega$, also are complete. So, it suffices to prove that for every $n \in \omega$ the topological group $Z_n(X)$ has precompact exponent. This will be proved by induction on n. For $n = 0$ the trivial group $Z_0(X) = \{e\}$ obviously has precompact exponent. Assume that for some $n \in \omega$ we have proved that the subgroup $Z_n(X)$ has precompact exponent. By Proposition 4, the center $Z(X/Z_n(X))$ of the quotient topological group $X/Z_n(X)$ has precompact exponent. Since $Z(X/Z_n(X)) = Z_{n+1}(X)/Z_n(X)$, we see that the quotient topological group $Z_{n+1}(X)/Z_n(X)$ has precompact exponent. By Proposition 2, the topological group $Z_{n+1}(X)$ has precompact exponent. □

Corollary 5 implies the following characterization of e:TS-closed nilpotent topological groups (announced in the introduction as Theorem 8).

Theorem 22. *For a nilpotent topological group X the following conditions are equivalent:*

(1)   X is e:TS-closed;
(2)   X is e:pTS-closed;
(3)   X is Weil-complete and has compact exponent;

(4)   X is complete and has compact exponent.

Proof. The implications $(2) \Rightarrow (1)$ and $(3) \Rightarrow (4)$ are trivial, and $(4) \Rightarrow (2)$ was proved in Theorem 21. It remains to prove that $(1) \Rightarrow (3)$. So, assume that the nilpotent topological group X is e:TS-closed. By Theorem 16, X is Weil-complete. By Corollary 5, for every $n \in \omega$ the subgroup $Z_n(X)$ has compact exponent. In particular, X has compact exponent, being equal to $Z_n(X)$ for a sufficiently large number n. □

We do not know if Theorem 22 remains true for hypercentral topological groups. We recall that a topological group X is *hypercentral* if for each closed normal subgroup $H \subsetneq X$ the quotient group X/H has non-trivial center. Each nilpotent topological group is hypercentral.

Problem 4. *Has each e:TS-closed hypercentral topological group compact exponent?*

The following characterization of compact topological groups shows that the e:pTS-closedness of X in Theorem 22 cannot be replaced by the e:sTS-closedness. The equivalence $(1) \Leftrightarrow (2)$ was proved by Gutik [3].

Theorem 23. *For a topological group X the following conditions are equivalent:*

(1)   X is compact;
(2)   X is e:sTS-closed;
(3)   X is e:rTG-closed.

Proof. The implication $(1) \Rightarrow (2, 3)$ is trivial.

To prove that $(2) \Rightarrow (1)$, assume that a topological group X is e:sTS-closed. Then it is e:TG-closed and hence complete (by Theorem 16). Assuming that X is not compact, we conclude that X is not totally bounded. So, there exists a neighborhood $V \subset X$ of the unit such that $X \not\subset FV \cup VF$ for any finite subset $F \subset X$.

Chose any element $0 \notin X$ and consider the space $X_0 = X \cup \{0\}$ endowed with the Hausdorff topology τ consisting of sets $W \subset X_0$ such that $W \cap X$ is open in X and if $0 \in W$, then $X \setminus W \subset FV$ for some finite subset $F \subset X$. Extend the group operation of X to a semigroup operation on X_0 letting $0x = 0 = x0$ for all $x \in X_0$. It is easy to see that X_0 is a Hausdorff semitopological semigroup containing X as a non-closed subgroup and witnessing that X is not e:sTS-closed.

To prove that $(3) \Rightarrow (1)$, assume that a topological group X is e:rTG-closed. Then it is e:TG-closed and hence complete (by Theorem 16). Assuming that X is not compact, we conclude that X is not totally bounded. By Lemma 2, X contains a countable subgroup which is not totally bounded. Now Lemma 3 (proved below) implies that X is not e:rTG-closed, which is a desired contradiction. □

A topology τ on a group X is called *right-invariant* (resp. *shift-invariant*) if $\{Ux : U \in \tau, x \in X\} = \tau$ (resp. $\{xUy : U \in \tau, x, y \in X\} = \tau$). This is equivalent to saying that (X, τ) is a right-topological (resp. semitopological) group.

Lemma 3. *If a topological group X contains a countable subgroup Z which is not totally bounded, then the group $\mathbb{Z} \times X$ admits a Hausdorff right-invariant topology τ such that the subgroup $\{0\} \times X$ is not closed in the right-topological group $(\mathbb{Z} \times X, \tau)$ and $\{0\} \times X$ is topologically isomorphic to X. Moreover, if the subgroup Z is central in X, then $(\mathbb{Z} \times X, \tau)$ is a semitopological group.*

Proof. Identify the product group $\mathbb{Z} \times X$ with the direct sum $\mathbb{Z} \oplus X$. In this case the group $X \subset \mathbb{Z} \oplus X$ is identified with the subgroup $\{0\} \times X$ of the group $\mathbb{Z} \times X$. Let $Z = \{z_k\}_{k \in \omega}$ be an enumeration of the countable subgroup Z. Since Z is not totally bounded, there exists a neighborhood $W = W^{-1} \subset X$ of the unit such that $Z \not\subset FW^3F$ for any finite subset $F \subset X$ (see Lemma 1). Using this property of Z, we can inductively construct a sequence of points $(x_n)_{n \in \omega}$ of Z such that for every $n \in \omega$ the following condition is satisfied:

(a) $x_n \notin F_n W^3 F_n^{-1}$ where

(b) $F_n = \{e\} \cup \{x_{i_1} x_{i_2} \cdots x_{i_k} z_j^\varepsilon : k \in \omega, \, n > i_1 > \cdots > i_k, \, j \le n, \, \varepsilon \in \{0, 1\}\}$.

For every $m \in \omega$ consider the subset

$$\Sigma_m := \{(0, e)\} \cup \{(n, x_{i_1} \cdots x_{i_n}) : n \in \mathbb{N}, \, i_1 > \cdots > i_n > m\} \subset \mathbb{Z} \times X.$$

On the group $G := \mathbb{Z} \oplus X$, consider the topology τ consisting of subsets $W \subset G$ such that for every $g \in W$ there exists $m \in \omega$ and a neighborhood $U_g \subset X$ of g such that $\Sigma_m U_g \subset W$. The definition of the topology τ implies that for any $W \in \tau$ and $a \in G$ the set Wa belongs to τ. So, (G, τ) is a right-topological group. If the subgroup Z is central, then for every $a \in G$ and $g \in aW$ we get $a^{-1}g \in W$, so we can find a neighborhood $U \subset X$ of $a^{-1}g$ and $m \in \omega$ such that $\Sigma_m U \subset W$. Then $U_g := aU$ is a neighborhood of g in X such that $\Sigma_m U_g = \Sigma_m a U = a \Sigma_m U \subset aW$, which means that the set aW belongs to the topology τ and the topology τ is invariant.

Let us show that for any open set $U \subset X$ and any $m \in \omega$ the set $\Sigma_m U$ belongs to the topology τ.

For every $g \in \Sigma_m U$ we can find $u \in U$ and a sequence $i_1 > \cdots i_n > m$ such that $g = x_{i_1} \cdots x_{i_n} u$. Choose neighborhoods $U_e, U_e' \subset X$ of the unit such that $uU_e \subset U$ and $U_e' x_{i_1} \cdots x_{i_n} u \subset x_{i_1} \cdots x_{i_n} u U_e$. Then

$$\Sigma_{i_1} U_e' g = \Sigma_{i_1} U_e' x_{i_1} \cdots x_{i_n} u \subset \Sigma_{i_1} x_{i_1} \cdots x_{i_n} u U_e \subset \Sigma_m U$$

and hence $\Sigma_m U \in \tau$.

Observe that for every $U \subset X$ and $m \in \omega$, have $\Sigma_m U \cap X = U$, which implies that X is a subgroup of the right-topological group $(\mathbb{Z} \oplus X, \tau)$. The subgroup X is not closed in $\mathbb{Z} \oplus X$ as $\bar{X}$ contains any point $(n, x) \in \mathbb{Z} \times X$ with $n \le 0$.

It remains to check that the right-topological semigroup (G, τ) is Hausdorff. Given any element $g = (n, x) \in G \setminus \{(0, e)\}$, we should find a neighborhood $U_e \subset X$ and $m \in \omega$ such that $\Sigma_m U_e \cap (\Sigma_m U_e g) = \emptyset$. If $x \notin \bar{Z}$, then we can find a neighborhood $U_e = U_e^{-1} \subset X$ of the unit such that $U_e x U_e \cap \bar{Z} = \emptyset$ and hence $\Sigma_0 U_e \cap (\Sigma_0 U_e g) \subset \mathbb{Z} \times (ZU_e \cap ZU_e x) = \emptyset$.

So, we assume that $x \in \bar{Z}$ and hence $x \in z_m W$ for some $m \in \omega$. Choose a neighborhood $V = V^{-1} \subset W$ of the unit such that $Vz_m \subset z_m W$ and if $x \ne e$, then $x \notin V^2$.

We claim that $\Sigma_m V \cap \Sigma_m V g = \emptyset$. Assuming that this intersection is not empty, fix an element $y \in \Sigma_m V \cap \Sigma_m V g$. The inclusion $y \in \Sigma_m V$ implies that $y = (k, x_{i_1} \cdots x_{i_k} v)$ for some numbers $k \in \mathbb{N}$, $i_1 > \cdots > i_k > m$, and $v \in V$. On the other hand, the inclusion $y \in \Sigma_m V g$ implies that $y = (l, x_{j_1} \cdots x_{j_l} u) g$ for some numbers $l \in \mathbb{N}$ and $j_1 > \cdots > j_l > m$ and some $u \in V$. It follows that

$$(k, x_{i_1} \cdots x_{i_k} v) = y = (l, x_{j_1} \cdots x_{j_l} u) \cdot (n, x) = (l + n, x_{j_1} \cdots x_{j_l} ux). \tag{1}$$

Let λ be the largest number $\le 1 + \min\{k, l\}$ such that $i_p = j_p$ for all $1 \le p < \lambda$. Three cases are possible.

(1) $\lambda \le \min\{k, l\}$. In this case the numbers i_λ and j_λ are well-defined and distinct. The Equality (1) implies $x_{i_1} \cdots x_{i_k} v = x_{j_1} \cdots x_{j_l} ux$. If $i_\lambda > j_\lambda$, then

$$x_{i_\lambda} = x_{j_\lambda} \cdots x_{j_l} uxv^{-1}(x_{i_{\lambda+1}} \cdots x_{i_k})^{-1} \subset x_{j_\lambda} \cdots x_{j_l} uz_m Wv^{-1}(x_{i_{\lambda+1}} \cdots x_{i_k})^{-1} \subset$$
$$\subset x_{j_\lambda} \cdots x_{j_l} Vz_m WV^{-1}(x_{i_{\lambda+1}} \cdots x_{i_k})^{-1} \subset x_{j_\lambda} \cdots x_{j_l} z_m WWV^{-1}(x_{i_{\lambda+1}} \cdots x_{i_k})^{-1} \subset F_{i_\lambda} W^3 F_{i_\lambda},$$

which contradicts the choice of x_{i_λ}.

If $i_\lambda < j_\lambda$, then

$$x_{j_\lambda} = x_{i_\lambda} \cdots x_{i_k} vx^{-1}u^{-1}(x_{j_{\lambda+1}} \cdots x_{j_l})^{-1} \subset x_{i_\lambda} \cdots x_{i_k} VW^{-1}z_m^{-1}V^{-1}(x_{j_{\lambda+1}} \cdots x_{j_l})^{-1} \subset$$
$$\subset x_{i_\lambda} \cdots x_{i_k} VW^{-1}W^{-1}z_m^{-1}(x_{j_{\lambda+1}} \cdots x_{j_l})^{-1} \subset x_{i_\lambda} \cdots x_{i_k} W^3(x_{j_{\lambda+1}} \cdots x_{j_l} z_m)^{-1} \subset F_{j_\lambda} W^3 F_{j_\lambda}^{-1},$$

which contradicts the choice of x_{j_λ}.

(2) $\lambda = 1 + \min\{k,l\}$ and $k = l$. In this case, Equation (1) implies that $n = 0$ and $v = ux$. Then $e \neq x = vu^{-1} \in V^2$, which contradicts the choice of V.

(3) $\lambda = 1 + \min\{k,l\}$ and $k < l$. In this case, Equation (1) implies that $v = x_{j_\lambda} \cdots x_{j_l} ux$ and hence $x_{j_\lambda} = vx^{-1}u^{-1}(x_{j_{\lambda+1}} \cdots x_{j_l})^{-1} \subset VW^{-1}z_m^{-1}V^{-1}(x_{j_{\lambda+1}} \cdots x_{j_l})^{-1} \subset VW^{-1}W^{-1}z_m^{-1}(x_{j_{\lambda+1}} \cdots x_{j_l})^{-1} \subset W^3(x_{j_{\lambda+1}} \cdots x_{j_l}z_m)^{-1} \subset F_{j_\lambda}W^3F_{j_\lambda}^{-1}$, which contradicts the choice of x_{j_λ}.

(4) $\lambda = 1 + \min\{k,l\}$ and $l < k$. In this case, Equation (1) implies that $x_{i_\lambda} \cdots x_{i_k} v = ux$ and hence $x_{i_\lambda} = uxv^{-1}(x_{i_{\lambda+1}} \cdots x_{i_k})^{-1} \subset Vz_mWV^{-1}(x_{i_{\lambda+1}} \cdots x_{i_k})^{-1} \subset z_mW^3(x_{i_{\lambda+1}} \cdots x_{i_k})^{-1} \subset F_{i_\lambda}W^3F_{i_\lambda}^{-1}$, which contradicts the choice of x_{i_λ}. This contradiction finishes the proof of the Hausdorff property of the topology τ. $\square$

Lemmas 2 and 3 have two implications.

Corollary 6. *Each e:sTG-closed topological group has compact center.*

Corollary 7. *An Abelian topological group is compact if and only if it is e:sTG-closed.*

Problem 5. *Is a topological group compact if it is e:sTG-closed?*

5. On i:$\mathcal{C}$-closed Topological Groups

In this section, we collect some results on i:$\mathcal{C}$-closed topological groups for various classes $\mathcal{C}$ of topologized semigroups. First we prove that for topological groups of precompact exponent, many of such closedness properties are equivalent.

Theorem 24. *For a topological group X of precompact exponent the following conditions are equivalent:*

(1)   *X is i:TS-closed;*
(2)   *X is i:TG-closed.*

Proof. The implications (1) $\Rightarrow$ (2) is trivial and follows from the inclusion TG $\subset$ TS. To prove that (2) $\Rightarrow$ (1), assume that X is i:TG-closed and take any continuous injective homomorphism $f : X \to Y$ to a Hausdorff topological semigroup Y. We need to show that $f(X)$ is closed in Y. Replacing Y by $\overline{f(X)}$, we can assume that the group $f(X)$ is dense in Y. We claim that Y is a topological group.

First observe that the image $e_Y = f(e_X)$ of the unit e_X of the group X is a two-sided unit of the semigroup Y (since the set $\{y \in Y : ye_Y = y = ye_Y\}$ is closed in Y and contains the dense subset $f(X)$).

Since the complete group X has precompact exponent, it has a compact exponent and hence for some number $n \in \mathbb{N}$ of the set $nX = \{x^n : x \in X\}$ has compact closure $K := \overline{nX}$. By the continuity of h and the Hausdorff property of Y, the image $f(K)$ is a compact closed subset of Y. Consequently, the set $Y_n = \{y \in Y : y^n \in f(K)\}$ is closed in Y. Taking into account that Y_n contains the dense subset $f(X)$, we conclude that $Y_n = Y$.

Now consider the compact subset $\Gamma = \{(x,y) \in f(K) \times f(K) : xy = e_Y\}$ in $Y \times Y$. Let $\mathrm{pr}_1, \mathrm{pr}_2 : \Gamma \to f(K)$ be the coordinate projections. We claim that these projections are bijective. Since $f(X)$ is a group, for every $z \in f(X)$ there exists a unique element $y \in f(X)$ with $zy = e_Y$. This implies that the projection $\mathrm{pr}_1 : \Gamma \to f(K)$, $\mathrm{pr}_1 : (z,y) \mapsto z$, is injective. Given any element $z \in f(K)$ find an element $x \in K \cap f^{-1}(z)$ and observe that $x^{-1} \in K^{-1} = K$ and hence the pair $(z,y) := (f(x), f(x^{-1}))$ belongs to Γ witnessing that the map $\mathrm{pr}_1 : \Gamma \to f(K)$ is surjective. Being a bijective continuous map defined on the compact space Γ, the map $\mathrm{pr}_1 : \Gamma \to f(K)$ is a homeomorphism. By analogy we can prove that the projection $\mathrm{pr}_2 : \Gamma \to f(K)$, $\mathrm{pr}_2 : (z,y) \mapsto y$, is a homeomorphism. Then the inversion map $i : f(K) \to f(K)$, $i = \mathrm{pr}_2 \circ \mathrm{pr}_1^{-1} : f(K) \to f(K)$ is continuous.

Now consider the continuous map $\tilde{i} : Y \to Y$ defined by $\tilde{i}(y) = y^{n-1} \cdot i(y^n)$ for $y \in Y$. This map is well-defined since $y^n \in f(K)$ for all $y \in Y$. Observe that for every element y of the group $f(X)$, the element $\bar{i}(y) = y^{n-1} \cdot i(y^n) = y^{n-1}(y^n)^{-1} = y^{-1}$ coincides with the inverse element of y in the group $f(X)$. Consequently, $y \cdot \bar{i}(y) = e_Y = \bar{i}_Y(y) \cdot y$ for all $y \in f(X)$ and by the continuity of the map $\bar{i}$ this equality holds for every $y \in Y$. This means that each element y of the semigroup Y has inverse $\bar{i}(y)$ and hence Y is a group. Moreover, the continuity of the map $\bar{i}$ ensures that Y is a topological group. So, $f : X \to Y$ is an injective continuous homomorphism to a topological group. Since X is i:TG-closed, the image $f(X)$ is closed in Y. $\square$

Theorems 22 and 24 imply the following characterization.

Corollary 8. *A nilpotent topological group X is* i:TS-*closed if and only if X is* e:TS-*closed and* i:TG-*closed.*

The above results allow us to reduce the problem of detecting i:TS-closed topological groups to the problem of detecting i:TG-closed topological groups. So, now we establish some properties of i:TG-closed topological groups.

Theorem 25. *The center of any* i:TG-*closed topological group X is compact.*

Proof. To derive a contradiction, assume that the center Z of an i:TG-closed topological group X is not compact. Being i:TG-closed, the topological group X is complete and so is its closed subsemigroup Z. By Theorem 3, the non-compact complete Abelian topological group Z is not i:TG-closed and hence admits a non-complete weaker Hausdorff group topology τ_Z.

Let $\mathcal{T}$ be the topology of X and $\mathcal{T}_e = \{U \in \mathcal{T} : e \in U\}$. Consider the family

$$\tau_e = \{V \cdot U : V \in \mathcal{T}_e, \; e \in U \in \tau_Z\}$$

of open neighborhoods of the unit in the topological group X. It can be shown that τ_e satisfies the Pontryagin Axioms ([27] 1.3.12) and hence is a base of some Hausdorff group topology $\tau \subset \mathcal{T}$ on X. Observe that the topology τ induces the topology τ_Z on the subgroup Z, which remains closed in the topology τ. Since the topological group (Z, τ_Z) is not complete, the topological group $X_\tau = (X, \tau)$ is not complete, too. Then the identity map $X \to \bar{X}_\tau$ into the completion $\bar{X}_\tau$ of X_τ has non-closed image, witnessing that the topological group X is not i:TG-closed. This is a desired contradiction, completing the proof of the theorem. $\square$

Theorem 25 can be reversed for connected nilpotent Lie groups.

Theorem 26. *A connected nilpotent Lie group X is* i:TG-*closed if and only if X has compact center.*

Proof. The "only if" part follows from Theorem 25 and the "if" part was proved by Omori ([4] Corollary 1.3) (see also [5] and [6] Theorem 5.1). $\square$

Example 1. An example of a non-compact connected nilpotent Lie group with compact center is the classical Weyl-Heisenberg group $H(w_0) := H(\mathbb{R})/Z$ where

$$H(\mathbb{R}) = \left\{ \begin{pmatrix} 1 & a & b \\ 0 & 1 & c \\ 0 & 0 & 1 \end{pmatrix} : a, b, c \in \mathbb{R} \right\} \quad \text{and} \quad Z = \left\{ \begin{pmatrix} 1 & 0 & b \\ 0 & 1 & 0 \\ 0 & 0 & 1 \end{pmatrix} : b \in \mathbb{Z} \right\}.$$

By Theorem 26, the Weyl-Heisenberg group $H(w_0)$ is i:TG-closed. On the other hand, $H(w_0)$ admits a continuous homomorphism onto the real line, which implies that $H(w_0)$ is not h:TG-closed. The group $H(w_0)$ is known to be minimal, see ([25] §5), ([34] 5.5), [35]. Being minimal and non-compact, the complete group $H(w_0)$ is not MAP.

We recall that a topological group X is *minimal* if each continuous bijective homomorphism $h : X \to Y$ to a topological group Y is a topological isomorphism. This definition implies the following (trivial) characterization.

Proposition 5. *A minimal topological group X is i:TG-closed if and only if X is e:TG-closed.*

Now we characterize i:TG-closed ω-narrow topological groups which are Čech-complete or Polish.

We recall that a topological group X is *ω-narrow* if for any neighborhood $U \subset X$ of the unit in X there exists a countable set $C \subset X$ such that $X = CU$. The following classical theorem of Guran [26] (see also ([27] Theorem 3.4.23)) describes the structure of ω-narrow topological groups.

Theorem 27 (Guran). *A topological group X is ω-narrow if and only if X is topologically isomorphic to a subgroup of a Tychonoff product $\prod_{\alpha \in A} P_\alpha$ of Polish groups.*

A topological group is called *Čech-complete* if its topological space is Čech-complete, i.e., is a G_δ-set in its Stone-Čech compactification. By Theorem 4.3.7 [27], each Čech-complete topological group is complete. By Theorem 4.3.20 [27], a topological group G is Čech-complete if and only if G contains a compact subgroup K such that the left quotient space $G/K = \{xK : x \in G\}$ is metrizable by a complete metric.

In the subsequent proofs we shall use the following known Open Mapping Principle for ω-narrow Čech-complete topological groups.

Theorem 28 (Open Mapping Principle, [27] Corollary 4.3.33). *Each continuous surjective homomorphism $h : X \to Y$ between ω-narrow Čech-complete topological groups is open.*

Theorem 29. *An ω-narrow i:TG-closed topological group X is Čech-complete if and only if it admits an injective continuous homomorphism $h : X \to Y$ to a Čech-complete topological group Y.*

Proof. The "only if" part is trivial. To prove the "if part", assume that X admits a continuous injective homomorphism $h : X \to Y$ to a Čech-complete topological group Y. Since X is i:TG-closed, the image $h(X)$ is closed in Y and hence $h(X)$ is a Čech-complete topological group. Replacing Y by $h(X)$, we can assume that $h(X) = Y$.

We claim that the bijective homomorphism $h : X \to Y$ is open and hence is a topological isomorphism. Given any open neighborhood $U \subset X$ of the unit, we should show that its image $h(U)$ is a neighborhood of the unit in Y. Using Guran's Theorem 27, we can find a continuous homomorphism $p : X \to P$ to a Polish group P and an open neighborhood $V \subset P$ of the unit such that $p^{-1}(V) \subset U$.

Consider the injective continuous homomorphism $ph : X \to P \times Y$, $ph : x \mapsto (p(x), h(x))$. Since the group X is ω-narrow and i:TG-closed, the image $ph(X)$ is an ω-narrow closed subgroup of the Čech-complete topological group $P \times Y$. Consequently, $ph(X)$ is an ω-narrow Čech-complete topological group. By Theorem 28, the projection pr $: ph(X) \to Y$, pr $: (x, y) \mapsto y$, is open. Consequently, $\mathrm{pr}(V) \subset h(U)$ is a neighborhood of the unit in Y. □

Problem 6. *Assume that X is an i:TG-closed ω-narrow topological group containing a compact G_δ-subgroup K. Is X Čech-complete?*

The answer to this problem is affirmative if the compact G_δ-subgroup $K \subset X$ is a singleton (in which case the topological group X has countable pseudocharacter).

We recall that a topological space X has *countable pseudocharacter* if for each point $x \in X$ there exists a countable family $\mathcal{U}$ of open sets in X such that $\bigcap \mathcal{U} = \{x\}$. By PG we denote the class of *Polish groups*, i.e., topological groups whose topological space is Polish (= separable completely metrizable).

Theorem 30. *An* i:PG-*closed topological group X is Polish if and only if X is ω-narrow and has countable pseudocharacter.*

Proof. The "only if" part is trivial. To prove the "if" part, assume that X is ω-narrow and has countable pseudocharacter. Using Guran's characterization of ω-narrow topological groups, we can show that X admits an injective continuous homomorphism $h : X \rightarrow Y$ into a Polish group Y. By analogy with the proof of Theorem 29, it can be shown that the homomorphism h is open and hence h is a topological isomorphism. So, X is Polish, being topologically isomorphic to the Polish group Y. $\square$

Now we present a characterization of i:TG-closed topological groups in the class of ω-narrow topological groups of countable pseudocharacter.

Theorem 31. *For an ω-narrow topological group X of countable pseudocharacter the following conditions are equivalent:*

(1)   X is Polish and minimal;
(2)   X is complete and minimal;
(3)   X is i:TG-*closed;*
(4)   X is i:PG-*closed.*

Proof. The implications $(1) \Rightarrow (2)$ and $(3) \Rightarrow (4)$ are trivial, and $(2) \Rightarrow (3)$ follows from (the trivial) Proposition 5. To prove that $(4) \Rightarrow (1)$, assume that the topological group X is i:PG-closed. By Theorem 30, the topological group X is Polish. To show that X is minimal, take any continuous bijective homomorphism $h : X \rightarrow Z$ to a topological group Z. Observe that the topological group Z is i:PG-closed, ω-narrow, and has countable pseudocharacter (being the continuous bijective image of the i:PG-closed Polish group X). By Theorem 30, the topological group Z is Polish and by Theorem 28, the homomorphism h is open and hence is a topological isomorphism. So, X is minimal. $\square$

Problem 7. *Is a topological group compact if it is* i:qTG-*closed?*

6. On h:$\mathcal{C}$-closed Topological Groups

In this section, we collect some results on h:$\mathcal{C}$-closed topological groups for various classes $\mathcal{C}$ of topologized semigroups. First observe the following trivial characterization of h:$\mathcal{C}$-closed topological groups.

Proposition 6. *Let $\mathcal{C}$ be a class of topological groups. A topological group X is* h:$\mathcal{C}$-*closed if and only if for any closed normal subgroup $N \subset X$ the quotient topological group X/N is* i:$\mathcal{C}$-*closed.*

Proposition 6 and Theorem 24 implies the following characterization.

Theorem 32. *For a topological group X of precompact exponent the following conditions are equivalent:*

(1)   X is h:TS-*closed;*
(2)   X is h:TG-*closed.*

Also Theorem 25 and Proposition 6 imply:

Corollary 9. *For any closed normal subgroup N of an* h:TG-*closed topological group X the quotient topological group X/N has compact center $Z(X/N)$. In particular, the Abelianization $X/\overline{[X,X]}$ of X is compact.*

We recall that for a topological group X its central series $\{e\} = Z_0(X) \subset Z_1(X) \subset \cdots$ consists of the subgroups defined recursively as $Z_{n+1}(G) = \{z \in X : \forall x \in X \ \ zxz^{-1}x^{-1} \in Z_n(X)\}$ for $n \in \omega$.

Proposition 7. *If a topological group X is h:TG-closed, then for every $n \in \omega$ the subgroup $Z_n(X)$ is compact.*

Proof. The compactness of the subgroups $Z_n(X)$ will be proved by induction on n. For $n = 0$ the compactness of the trivial group $Z_0(X) = \{e\}$ is obvious. Assume that for some $n \in \omega$ we have proved that the subgroup $Z_n(X)$ is compact. By Corollary 9, the center $Z(X/Z_n(X))$ of the quotient topological group $X/Z_n(X)$ is compact. Since $Z(X/Z_n(X)) = Z_{n+1}(X)/Z_n(X)$, we see that the quotient topological group $Z_{n+1}(X)/Z_n(X)$ is compact. By ([27] Corollary 1.5.8) (the 3-space property of the compactness), the topological group $Z_{n+1}(X)$ is compact. $\square$

Proposition 7 implies the following characterization of h:TG-closed nilpotent topological groups, first proved by Dikranjan and Uspenskij ([10] 3.9).

Corollary 10 (Dikranjan, Uspenskij). *A nilpotent topological group is compact if and only if it is h:TG-closed.*

We recall that a topological group X is *totally minimal* if for any closed normal subgroup $N \subset X$ the quotient topological group X/N is minimal. Proposition 6 and Theorem 31 imply the following characterization.

Corollary 11. *For an ω-narrow topological group X of countable pseudocharacter the following conditions are equivalent:*

(1)   *X is h:TG-closed;*
(2)   *X is h:PG-closed.*
(3)   *X is Polish and totally minimal.*

In fact, the countable pseudocharacter can be removed from this corollary. Following [11], we define a topological group X to be *totally complete* if for any closed normal subgroup $N \subset X$ the quotient topological group X/N is complete. It is easy to see that each h:TG-closed topological group is totally complete. By ([11] 3.45), a topological group is h:TG-closed if it is totally complete and totally minimal. These observations are complemented by the following characterization.

Theorem 33. *For an ω-narrow topological group X the following conditions are equivalent:*

(1)   *X is h:TG-closed;*
(2)   *X is totally complete and totally minimal.*

Proof. The implication (2) $\Rightarrow$ (1) was proved by Lukács ([11] 3.45). The implication (1) $\Rightarrow$ (2) will be derived from the following lemma.

Lemma 4. *Each ω-narrow h:TG-closed topological group X is minimal.*

Proof. We should prove that each continuous bijective homomorphism $f : X \to Y$ to a topological group Y has continuous inverse $f^{-1} : Y \to X$. Since X embeds into the Tychonoff product of Polish groups, it suffices to check that for every continuous homomorphism $p : X \to P$ to a Polish group P the composition $p \circ f^{-1} : Y \to P$ is continuous. Since X is h:TG-closed, the image $p(X)$ is closed in P. Replacing P by $p(X)$, we can assume that $P = p(X)$. Let $K_X = p^{-1}(1)$ be the kernel of the homomorphism p. Observe that the quotient topological group X/K_X admits a bijective continuous homomorphism $\bar{p} : X/K_X \to P$ such that $p = \bar{p} \circ q_X$ where $q_X : X \to X/K_X$ is the quotient homomorphism. It follows that the quotient topological group X/K_X has countable pseudocharacter. Moreover, the topological group X/K_X is ω-narrow and h:TG-closed being a continuous homomorphic image of the ω-narrow h:TG-closed topological group X. By Corollary 11, the group X/K_X is Polish and minimal. Consequently, $\bar{p} : X/K_X \to P$ is a topological isomorphism.

Now we shall prove that the image $K_Y = f(K_X)$ of K_X is closed in Y. Consider the continuous homomorphism $fp : X \to Y \times P$, $fp : x \mapsto (f(x), p(x))$, and observe that its image $fp(X)$ is a closed subgroup of $Y \times P$ by the i:TG-closedness of X. Consequently, the homomorphism $p \circ f^{-1}$ has closed graph $\Gamma = \{(y, p \circ f^{-1}(y)) : y \in Y\} = \{(f(x), p(x)) : x \in X\} = fp(X)$. Since the intersection $\Gamma \cap (Y \times \{1\})$ is a closed subset of $Y \times \{1\}$ the homomorphism $p \circ f^{-1}$ has closed kernel

$$K_Y = \{y \in Y : p \circ f^{-1}(y) = 1\} = \{y \in Y : (y, 1) \in \Gamma\},$$

which coincides with $f(K_X)$. Let Y/K_Y be the quotient topological group and $q_Y : Y \to Y/K_Y$ be the quotient homomorphism.

The continuous bijective homomorphism $f : X \to Y$ induces a continuous bijective homomorphism $\hat{f} : X/K_X \to Y/K_Y$ making the following diagram commutative.

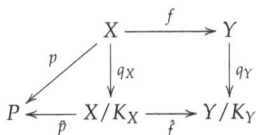

$$
\begin{array}{ccc}
X & \xrightarrow{\ f\ } & Y \\
{\scriptstyle p}\swarrow \ \ \downarrow{\scriptstyle q_X} & & \downarrow{\scriptstyle q_Y} \\
P \xleftarrow[\bar{p}]{} X/K_X & \xrightarrow[\hat{f}]{} & Y/K_Y
\end{array}
$$

The minimality of the Polish group X/K_X guarantees that the bijective homomorphism $\hat{f}$ is a topological isomorphism, which implies that the homomorphism $\bar{p} \circ \hat{f}^{-1} \circ q_Y = p \circ f^{-1} : Y \to P$ is continuous. $\square$

Now we are able to prove the implication $(1) \Rightarrow (2)$ of Theorem 33. Given an ω-narrow h:TG-closed topological group X and a closed normal subgroup $N \subset X$, we should check that the quotient topological group X/N is complete and minimal. Observe that X/N is ω-narrow and h:TG-closed (being a continuous homomorphic image of the ω-bounded h:TG-closed topological group X). By Theorem 16, the h:TG-closed topological group X/N is complete and by Lemma 4, it is minimal. $\square$

Remark 2. *By Theorem 5.1 of [6], the class of* h:TG-*closed topological groups includes all quasi semi-simple topological groups (introduced in Definition 3.2 of [6]). Applying Theorem 33, we conclude that each ω-narrow quasi semi-simple topological group is totally minimal, which generalizes Corollary 5.3 of [6] (establishing the total minimality of separable quasi semi-simple groups).*

7. On p:$\mathcal{C}$-closed Topological Groups

In this section, we collect some results on p:$\mathcal{C}$-closed topological groups for various classes $\mathcal{C}$ of topologized semigroups.

Theorem 34. *Let $\mathcal{C}$ be a class of topologized semigroups, containing all Abelian topological groups. For a topological group X the following conditions are equivalent:*

(1) *X is* p:$\mathcal{C}$-*closed;*
(2) *each closed subsemigroup of X is* h:$\mathcal{C}$-*closed;*
(3) *each closed subgroup of X is* h:$\mathcal{C}$-*closed.*

Proof. The equivalence $(1) \Leftrightarrow (2)$ follows immediately from the definitions of p:$\mathcal{C}$-closed and h:$\mathcal{C}$-closed topological groups, and $(2) \Rightarrow (3)$ is trivial.

To prove that $(3) \Rightarrow (2)$, assume that each closed subgroup of X is h:$\mathcal{C}$-closed, and take any closed subsemigroup $S \subset X$ of X. We claim that S is a subgroup of X. Given any element $x \in S$, consider the closure Z of the cyclic subgroup $\{x^n\}_{n \in \mathbb{Z}}$. By our assumption, Z is h:$\mathcal{C}$-closed and by Theorem 3, the Abelian h:$\mathcal{C}$-closed topological group Z is compact. Being a compact monothetic group, Z coincides with the closure of the subsemigroup $\{x^n\}_{n\in\mathbb{N}}$ and hence is contained in the closed subsemigroup S. Then $x^{-1} \in Z \subset S$ and S is a subgroup of X. By (3), S is h:$\mathcal{C}$-closed. $\square$

For topological groups of precompact exponent, Theorem 32 implies the following characterization.

Corollary 12. *For any topological group X of precompact exponent the following conditions are equivalent:*

(1) X is p:TG-*closed*;
(2) X is p:TS-*closed*.

An interesting property of p:TG-closed topological groups was discovered by Dikranjan and Uspenskij ([10] 3.10). This property concerns the transfinite derived series $(X^{[\alpha]})_\alpha$ of a topological group X, which consists of the closed normal subgroups $X^{[\alpha]}$ of X, defined by the transfinite recursive formulas:

$$X^{[0]} := X,$$

$$X^{[\alpha+1]} := \overline{[X^{[\alpha]}, X^{[\alpha]}]} \text{ for each ordinal } \alpha,$$

$$X^{[\gamma]} := \bigcap_{\alpha<\gamma} X^{[\alpha]} \text{ for each limit ordinal } \gamma.$$

The closed subgroup

$$X^{[\infty]} := \bigcap_\alpha X^{[\alpha]}$$

is called the *hypocommutator* of X. A topological group is called *hypoabelian* it its hypocommutator is trivial.

Theorem 35 (Dikranjan, Uspenskij). *For each p:TG-closed topological group X the quotient topological group $X/X^{[\infty]}$ is compact.*

Corollary 13. *A hypoabelian topological group is compact if and only if it is p:TG-closed.*

Since solvable topological groups are hypoabelian, Corollary 13 implies the following characterization of compactness of solvable topological groups.

Corollary 14. *A solvable topological group X is compact if and only if it is p:TG-closed.*

By ([10] 2.16), *a balanced topological group is p:TG-closed if and only if it is c:TG-closed.*

Problem 8. *Is each balanced p:TS-closed topological group c:TS-closed?*

8. On Closedness Properties of MAP Topological Groups

In this section, we establish some properties of MAP topological groups. We recall that a topological group X is *maximally almost periodic* (briefly, *MAP*) if it admits a continuous injective homomorphism into a compact topological group.

Theorem 36. *A topological group X is compact if and only if X is p:TG-closed and MAP.*

Proof. The "only if" part is trivial. To prove the "if' part, assume that X is p:TG-closed and MAP. Then X is complete. Assuming that X is not compact, we can apply Lemma 2 and find a non-compact closed separable subgroup $H \subset X$. Since X is p:TG-closed, the closed subgroup H of X is h:TG-closed and by Lemma 4, it is minimal and being complete and MAP, is compact. However, this contradicts the choice of H. $\square$

The notion of a MAP topological group can be generalized as follows. Let $\mathcal{K}$ be a class of topologized semigroups. A topologized semigroup X is defined to be $\mathcal{K}$-MAP if it admits a continuous injective homomorphism $f : X \to K$ to some compact topologized semigroup $K \in \mathcal{K}$. So, MAP is equivalent to TG-MAP.

Theorem 37. *Let $\mathcal{C}, \mathcal{K}$ be two classes of Hausdorff topologized semigroups such that for any $C \in \mathcal{C}$ and $K \in \mathcal{K}$ the product $C \times K$ belongs to the class $\mathcal{C}$. A $\mathcal{K}$-MAP topologized semigroup X is i:$\mathcal{C}$-closed if and only if it is h:$\mathcal{C}$-closed.*

Proof. The "if" part is trivial. To prove the "only if" part, assume that a topologized group X is $\mathcal{K}$-MAP and i:$\mathcal{C}$-closed. To prove that X is h:$\mathcal{C}$-closed, take any continuous homomorphism $f : X \to Y$ to a topologized semigroup $Y \in \mathcal{C}$. Since X is $\mathcal{K}$-MAP, there exists a continuous injective homomorphism $h : X \to K$ into a compact topologized semigroup $K \in \mathcal{K}$. By our assumption, the topologized semigroup $Y \times K$ belongs to the class $\mathcal{C}$. Since the homomorphism $fh : X \to Y \times K$, $fh : x \mapsto (f(x), h(x))$, is continuous and injective, the image $\Gamma = fh(X)$ of the i:$\mathcal{C}$-closed semigroup X in the semigroup $Y \times K \in \mathcal{C}$ is closed. By ([31] 3.7.1), the projection $\mathrm{pr} : Y \times K \to Y$ is a closed map (because of the compactness of K). Then the projection $\mathrm{pr}(fh(X)) = h(X)$ is closed in Y. $\square$

Since compact topological groups are balanced, each MAP topological group admits a continuous injective homomorphism into a balanced topological group. By [27] (p. 69) a topological group X is balanced iff each neighborhood $U \subset X$ of the unit contains a neighborhood $V \subset X$ of the unit which is *invariant* in the sense that $V = xVx^{-1}$ for all $x \in X$.

Proposition 8. *Let X be an i:TG-closed topological group. For each continuous injective homomorphism $f : X \to Y$ to a balanced topological group Y and each closed normal subgroup $Z \subset X$ the image $f(Z)$ is closed in Y.*

Proof. To derive a contradiction, assume that the image $f(Z)$ of some closed normal subgroup $Z \subset X$ is not closed in Y. Let $\mathcal{B}_X$ be the family of open neighborhoods of the unit in the topological group X and $\mathcal{B}_Y$ be the family of open invariant neighborhoods of the unit in the balanced topological group Y. It can be shown that the family

$$\mathcal{B} = \{V \cdot (f|Z)^{-1}(W) : V \in \mathcal{B}_X, \ W \in \mathcal{B}_Y\}$$

satisfies the Pontryagin Axioms ([27] 1.3.12) and hence is a base of some Hausdorff group topology τ on X. In this topology the subgroup Z is closed and is topologically isomorphic to the topological group $f(Z)$ which is not closed in Y and hence is not complete. Then the topological group $X_\tau = (X, \tau)$ is not complete too, and hence is not closed in its completion $\bar{X}_\tau$. Now we see that the identity homomorphism $\mathrm{id} : X \to X_\tau \subsetneq \bar{X}_\tau$ witnesses that X is not i:TG-closed. This contradiction completes the proof. $\square$

In the proof of our next result, we shall need the (known) generalization of the Open Mapping Principle 28 to homomorphisms between K-analytic topological groups.

We recall [36] that a topological space X is *K-analytic* if $X = f(Z)$ for some continuous function $f : Z \to X$ defined on an $F_{\sigma\delta}$-subset Z of a compact Hausdorff space C. It is clear that the continuous image of a K-analytic space is K-analytic and the product $A \times C$ of a K-analytic space A and a compact Hausdorff space C is K-analytic.

A topological group is called *K-analytic* if its topological space is K-analytic. In ([36] §I.2.10) it was shown that Open Mapping Principle 28 generalizes to homomorphisms defined on K-analytic groups.

Theorem 38 (K-analytic Open Mapping Principle, ([36] §I.2.10)). *Each continuous homomorphism $h : X \to Y$ from a K-analytic topological group X onto a Baire topological group Y is open.*

Theorem 38 will be used in the proof of the following lemma.

Lemma 5. *Let X be an* i:TG-*closed MAP topological group. For a closed normal subgroup* $Z \subset X$ *and a continuous homomorphism* $h : X \to Y$ *to a topological group Y, the image* $h(Z)$ *is compact if and only if* $h(Z)$ *is contained in a K-analytic subspace of Y.*

Proof. The "only if" part is trivial. To prove the "if" part, assume that the image $h(Z)$ is contained in a K-analytic subspace A of Y.

Being MAP, the group X admits a continuous injective homomorphism $f : X \to K$ to a compact topological group K. By Proposition 8, the image $f(Z)$ is a compact subgroup of K.

Now consider the continuous injective homomorphism $fh : X \to K \times Y$, $fh : x \mapsto (f(x), h(x))$. By the i:TG-closedness, the image $\Gamma := fh(X)$ is closed in $K \times Y$. Then the space $H = \Gamma \cap (f(Z) \times A)$ is K-analytic (as a closed subspace of the K-analytic space $K \times A$). Observe that $H = \{(f(z), h(z) : z \in Z\}$ is a subgroup of $X \times Y$.

By the K-analytic Open Mapping Principle 38, the bijective continuous homomorphism $\mathrm{pr}_K : H \to f(Z)$, $\mathrm{pr}_K : (x, y) \mapsto x$, from the K-analytic group H to the compact group $f(Z)$ is open and hence is a topological isomorphism. Consequently, the topological group H is compact and so is its projection $h(Z)$ onto Y. □

We recall that a topological group is *ω-balanced* iff it embeds into a Tychonoff product of metrizable topological groups.

Corollary 15. *If an ω-balanced MAP topological group X is* i:TG-*closed, then each ω-narrow closed normal subgroup of X is compact.*

Proof. Being ω-balanced and complete, the topological group X can be identified with a closed subgroup of a Tychonoff product $\prod_{\alpha \in A} M_\alpha$ of complete metrizable topological groups.

Fix an ω-narrow closed normal subgroup H in X and observe that for every $\alpha \in A$ the projection $\mathrm{pr}_\alpha(H) \subset M_\alpha$ is an ω-narrow and hence separable subgroup of the metrizable topological group M_α. Then the closure of $\mathrm{pr}_\alpha(H)$ in the complete metrizable topological group M_α is a Polish (and hence K-analytic) group. By Proposition 8, the group $\mathrm{pr}_\alpha(H)$ is compact. Being a closed subgroup of the product $\prod_{\alpha \in A} \mathrm{pr}_\alpha(H)$ of compact topological groups, the topological group H is compact, too. □

We recall that for a topological group X the *ω-conjucenter* $Z^\omega(X)$ of X consists of the points $z \in X$ whose conjugacy class $C_X(z) := \{xzx^{-1} : x \in X\}$ is ω-narrow in X. A subset A of a topological group X is called *ω-narrow* if for each neighborhood $U \subset X$ of the unit there exists a countable set $B \subset X$ such that $A \subset BU \cap UB$.

Theorem 39. *Each* i:TG-*closed ω-balanced MAP topological group X has compact ω-conjucenter $Z^\omega(X)$.*

Proof. First we prove that $Z^\omega(X)$ is precompact. Assuming the opposite, we can apply Lemma 2 and find a countable subgroup $D \subset Z^\omega(X)$ whose closure $\bar{D}$ is not compact. By the definition of $Z^\omega(X)$, each element $x \in D$ has ω-narrow conjugacy class $C_X(x)$. By ([27] 5.1.19), the ω-narrow set $\bigcup_{x \in D} C_X(x)$ generates an ω-narrow subgroup H. It is clear that H is normal. By Corollary 15, the closure $\bar{H}$ of H is compact, which is not possible as $\bar{H}$ contains the non-compact subgroup $\bar{D}$. This contradiction completes the proof of the precompactness of $Z^\omega(X)$. Then the closure $\bar{Z}^\omega(X)$ of the subgroup $Z^\omega(X)$ in X is a compact normal subgroup of X. The normality of $\bar{Z}^\omega(X)$ guarantees that for every $z \in \bar{Z}^\omega(X)$ the conjugacy class $C_X(z) \subset \bar{Z}^\omega(X)$ is precompact and hence ω-narrow, which means that $z \in Z^\omega(X)$. Therefore, the ω-conjucenter $Z^\omega(X) = \bar{Z}^\omega(X)$ is compact. □

For any topological group X let us define an increasing transfinite sequence $(Z_\alpha(X))_\alpha$ of closed normal subgroups defined by the recursive formulas

$$Z_0(X) = \{e\},$$
$$Z_{\alpha+1}(X) := \{z \in X : \forall x \in X \; xzxz^{-1} \in Z_\alpha(X)\} \text{ for any ordinal } \alpha \text{ and}$$
$$Z_\beta(X) \text{ is the closure of the normal group } \bigcup_{\alpha<\beta} Z_\beta(X) \text{ for a limit ordinal } \beta.$$

The closed normal subgroup $Z_\infty(X) = \bigcup_\alpha Z_\alpha(X)$ is called the *hypercenter* of the topological group X. We recall that a topological group X is *hypercentral* if for every closed normal subgroup $N \subsetneq X$ the quotient topological group X/N has non-trivial center $Z(X/N)$. It is easy to see that a topological group X is hypercentral if and only if its hypercenter equals X. It follows that a discrete topological group is hypercentral if and only if it is hypercentral in the standard algebraic sense ([37] 364). Observe that each nilpotent topological group is hypercentral. More precisely, a group X is nilpotent if and only if $Z_n(X) = X$ for some finite number $n \in \mathbb{N}$.

Corollary 16. *If an ω-balanced MAP topological group X is* i:TG-*closed, then its hypercenter $Z_\infty(X)$ is contained in the ω-conjucenter $Z^\omega(X)$ and hence is compact.*

Proof. By Theorem 39, the ω-conjucenter $Z^\omega(X)$ of X is compact. By transfinite induction we shall prove that for every ordinal α the subgroup $Z_\alpha(X)$ is contained in $Z^\omega(X)$. This is trivial for $\alpha = 0$. Assume that for some ordinal α we have proved that $\bigcup_{\beta<\alpha} Z_\beta(X) \subset Z^\omega(X)$. If the ordinal α is limit, then

$$Z_\alpha(X) = \overline{\bigcup_{\beta<\alpha} Z_\beta(X)} \subset Z^\omega(X).$$

Next, assume that $\alpha = \beta + 1$ is a successor ordinal. To prove that $Z_\alpha(X) \subset Z^\omega(X)$, take any point $z \in Z_\alpha(X)$ and observe that $C_X(z) = \{xzx^{-1} : x \in X\} \subset Z_\beta(X) \subset Z^\omega(X)$ and hence $C_X(z)$ is ω-narrow. So, $z \in Z^\omega(X)$ and $Z_\alpha(X) \subset Z^\omega(X)$.

By the Principle of Transfinite Induction, the subgroup $Z_\alpha(X) \subset Z^\omega(X)$ for every ordinal α. Then the hypercenter $Z_\infty(X)$ is contained in $Z^\omega(X)$ and is compact, being equal to $Z_\alpha(X)$ for a sufficiently large ordinal α. $\square$

Corollary 17. *A hypercentral topological group X is compact if and only if X is* i:TG-*closed, ω-balanced, and MAP.*

Problem 9. *Is each* h:TG-*closed hypercentral MAP topological group compact?*

9. The Compactness of h:TG-closed MAP-Solvable Topological Groups

In this section, we detect compact topological groups among h:TG-closed MAP-solvable topological groups. We define a topological group X to be *MAP-solvable* if there exists a decreasing sequence $X = X_0 \supset X_1 \supset \cdots \supset X_m = \{e\}$ of closed normal subgroups in X such that for every $n < m$ the quotient group X_n/X_{n+1} is Abelian and MAP. It is clear that each MAP-solvable topological group is solvable. By the Pontryagin Duality ([27] 9.7.5) (see also [38]), each locally compact Abelian group is MAP. This observation implies the following characterization.

Proposition 9. *A locally compact topological group is MAP-solvable if and only if it is solvable.*

Now we prove that balanced MAP-solvable h:TG-closed topological groups are compact. We recall that a topological group X is called a *balanced* if each neighborhood $U \subset X$ of the unit contains a neighborhood $V \subset X$ of the unit such that $xVx^{-1} \subset U$ for all $x \in X$.

Theorem 40. *For a solvable topological group X the following conditions are equivalent:*

(1) *X is compact;*
(2) *X is balanced, locally compact and h:TG-closed;*
(3) *X is balanced, MAP-solvable and h:TG-closed.*

Proof. The implication (1) $\Rightarrow$ (2) is trivial and (2) $\Rightarrow$ (3) follows from Proposition 9. To prove that (3) $\Rightarrow$ (1), assume that a topological group X balanced, MAP-solvable and h:TG-closed. Being MAP-solvable, X admits a decreasing sequence of closed normal subgroups $X = X_0 \supset X_2 \supset \cdots \supset X_m = \{e\}$ such that for every $n < m$ the quotient group X_n/X_{n+1} is Abelian and MAP.

To prove that the group $X = X/X_m$ is compact, it suffices to show that for every $n \leq m$ the quotient group X/X_n is compact. This is trivial for $n = 0$. Suppose that for some $n < m$ the group X/X_n is compact. Assuming that the quotient group $G := X/X_{n+1}$ is not compact, we conclude that the normal Abelian subgroup $A := X_n/X_{n+1}$ of G is not compact. The h:TG-closedness of X implies the h:TG-closedness of the quotient group G. Then G is complete and by Lemma 2, the non-compact closed subgroup A of X contains a countable subgroup Z whose closure $\bar{Z}$ is not compact.

Claim 2. *For every $a \in A$ its conjugacy class $C_G(a) = \{xax^{-1} : x \in G\}$ is compact.*

Proof. Consider the continuous map $f : G \to A$, $f : x \mapsto xax^{-1}$, and observe that it is constant on cosets xA, $x \in G$. Consequently, there exists a function $\check{f} : G/A \to A$ such that $f = \check{f} \circ q$ where $q : G \to G/A$ is a quotient homomorphism. Since the group G/A carries the quotient topology, the continuity of f implies the continuity of $\check{f}$. Now the compactness of the quotient group $G/A = X/X_n$ implies that the set $\check{f}(G/A) = f(G) = C_G(a)$ is compact. $\square$

It follows that the union $\bigcup_{z \in Z} C_G(z)$ is σ-compact and hence generates a σ-compact subgroup $H \subset A$, which is normal in G. Then its closure $\bar{H}$ is a normal closed ω-narrow subgroup in G.

Claim 3. *The topological group G is balanced and MAP.*

Proof. The topological group G is balanced, being a quotient group of the balanced topological group X. Let $\mathcal{B}_G$ be the family of all open invariant neighborhoods of the unit in the balanced group G. To prove that X is MAP, we shall use the fact that the Abelian topological group A is MAP, which implies that the strongest totally bounded group topology τ_A on A is Hausdorff. Let us observe that for every $x \in G$ the continuous automorphism $A \to A$, $a \mapsto xax^{-1}$, remains continuous in the topology τ_A. Using this fact, it can be shown that the family

$$\mathcal{B} = \{UV : U \in \mathcal{B}_G, \ V \in \tau_A\}$$

satisfies the Pontryagin Axioms ([27] 1.3.12) and hence is a base of some Hausdorff group topology τ on G. Observe that the subgroup A of (G, τ) is precompact and has compact quotient group $(G, \tau)/A$. Since the precompactness is a 3-space property (see [27] 1.5.8), the topological group (G, τ) is precompact and its Raikov-completion $\bar{G}$ is compact. The identity homomorphism id : $G \to (G, \tau) \subset \bar{G}$ witnesses that the topological group G is MAP. $\square$

Since the topological group G is balanced, MAP and h:TG-closed, we can apply Corollary 15 and conclude that the ω-narrow closed normal subgroup H of G is compact, which is not possible as H contains a closed non-compact subgroup $\bar{Z}$. This contradiction completes the proof of the compactness of the subgroup $A = X_n/X_{n+1}$. Now the compactness of the groups X/X_n and $A = X_n/X_{n+1}$ imply the compactness of the quotient group X/X_{n+1}, see ([27] 1.5.8). $\square$

Since each discrete topological group is locally compact and balanced, Theorem 40 implies

Corollary 18. *A solvable topological group is finite if and only if it is discrete and h:TG-closed.*

10. Some Counterexamples

In this section, we collect some counterexamples.

Our first example shows that Theorem 22 does not generalize to solvable (even meta-Abelian) discrete groups. A group G is called *meta-Abelian* if it contains a normal Abelian subgroup H with Abelian quotient G/H.

For an Abelian group X let $X \rtimes C_2$ be the product $X \times \{-1, 1\}$ endowed with the operation $(x, u) * (y, v) = (xy^u, uv)$ for $(x, u), (y, v) \in X \times \{-1, 1\}$. The semidirect product $X \rtimes C_2$ is meta-Abelian (since $X \times \{1\}$ is a normal Abelian subgroup of index 2 in $X \rtimes C_2$).

By T_1S we denote the family of topological semigroups X satisfying the separation axiom T_1 (which means that finite subsets in X are closed). Since $TS \subset T_1S$, each T_1S-closed topological semigroup is TS-closed.

Proposition 10. *For any Abelian group X the semidirect product $X \rtimes C_2$ endowed with the discrete topology is an e:T_1S-closed MAP topological group.*

Proof. First we show that the group $X \rtimes C_2$ is MAP. By Pontryagin Duality [27] Theorem 9.7.5, the Abelian discrete topological group X is MAP and hence admits an injective homomorphism $\delta : X \to K$ to a compact Abelian topological group K. It easy to see that the semidirect product $K \rtimes C_2$ endowed with the group operation $(x, u) * (y, v) = (xy^u, uv)$ for $(x, u), (y, v) \in K \times C_2$ is a compact topological group and the map $\delta_2 : X \rtimes C_2 \to K \rtimes C_2$, $\delta_2 : (x, u) \mapsto (\delta(x), u)$, is an injective homomorphism witnessing that the discrete topological group $X \rtimes C_2$ is MAP.

To show that $X \rtimes C_2$ is e:T_1S-closed, fix a topological semigroup $Y \in T_1S$ containing the group $G := X \rtimes C_2$ as a discrete subsemigroup. Identify X with the normal subgroup $X \times \{1\}$ of G. First we show that X is closed in Y. Assuming the opposite, we can find an element $\hat{y} \in \bar{X} \setminus X$. Consider the element $p := (e, -1) \in X \rtimes C_2$ and observe that for any element $x \in X$ we get $pxp = (e, -1)(x, 1)(e, -1) = (x^{-1}, 1) = x^{-1}$ and hence $pxpx = e$, where e is the unit of the groups $X \subset G$. The closedness of the singleton $\{e\}$ in Y and the continuity of the multiplication in the semigroup Y guarantee that the set

$$Z = \{y \in Y : ypyp = e\}$$

is closed in Y and hence contains the closure of the group X in Y. In particular, $\hat{y} \in \bar{X} \subset \bar{Z} = Z$. So, $\hat{y}p\hat{y}p = e$. Since the subgroup G of Y is discrete, there exists a neighborhood $V_e \subset Y$ of e such that $V_e \cap G = \{e\}$. By the continuity of the semigroup operation on Y, the point $\hat{y}$ has a neighborhood $W \subset Y$ such that $WpWp \subset V_e$. Fix any element $x \in W \cap X$ and observe that $(W \cap X)pxp \subset X \cap (WpWp) \subset X \cap V_e = \{e\}$, which is not possible as the set $W \cap X$ is infinite and so is its right shift $(W \cap X)pxp$ in the group G. This contradiction shows that the set X is closed in Y.

Next, we show that the shift $Xp = X \times \{-1\}$ of the set X is closed in Y. Assuming that Xp has an accumulating point y^* in Y, we conclude that y^*p is an accumulating point of the group X, which is not possible as X is closed in Y. So, the sets X and Xp are closed in Y and so is their union $G = X \cup Xp$, witnessing that the group G is e:T_1S-closed. $\square$

Since the isometry group $\mathrm{Iso}(\mathbb{Z})$ of $\mathbb{Z}$ is isomorphic to the semidirect product $\mathbb{Z} \rtimes C_2$, Proposition 10 implies the following fact.

Example 2. *The group $\mathrm{Iso}(\mathbb{Z})$ endowed with the discrete topology is e:T_1S-closed and MAP but does not have compact exponent.*

By Dikranjan-Uspenskij Theorem 9, each h:TG-closed nilpotent topological group is compact. Our next example shows that this theorem does not generalize to solvable topological groups and thus resolves in negative Question 3.13 in [10] and Question 36 in [12].

Example 3. *The Lie group*

$$\mathrm{Iso}_+(\mathbb{C}) = \left\{ \begin{pmatrix} a & b \\ 0 & 1 \end{pmatrix} : a, b \in \mathbb{C}, \ |a| = 1 \right\}$$

of orientation-preserving isometries of the complex plane is h:TS-closed and MAP-solvable but not compact and not MAP.

Proof. The group $\mathrm{Iso}_+(\mathbb{C})$ is topologically isomorphic to the semidirect product $\mathbb{C} \rtimes \mathbb{T}$ of the additive group $\mathbb{C}$ of complex numbers and the multiplicative group $\mathbb{T} = \{z \in \mathbb{C} : |z| = 1\}$. The semidirect product is endowed with the group operation $(x, a) \cdot (y, b) = (x + ay, ab)$ for $(x, a), (y, b) \in \mathbb{C} \rtimes \mathbb{T}$.

It is clear that the group $G := \mathbb{C} \rtimes \mathbb{T}$ is meta-Abelian (and hence solvable) and not compact. Being solvable and locally compact, the group G is MAP-solvable (see Proposition 9). To prove that G is h:TS-closed, take any continuous homomorphism $h : G \to Y$ to a Hausdorff topological semigroup Y and assume that the image $h(G)$ is not closed in Y. Replacing Y by $\overline{h(G)}$, we can assume that the subgroup $h(G)$ is dense in the topological semigroup Y.

Claim 4. *The homomorphism h is injective.*

Proof. Consider the closed normal subgroup $H = h^{-1}(h(0,1))$ of G, the quotient topological group G/H and the quotient homomorphism $q : G \to G/H$. It follows that $h = \tilde{h} \circ q$ for some continuous homomorphism $\tilde{h} : G/H \to Y$. We claim that the subgroup H is trivial. To derive a contradiction, assume that H contains some element $(x, a) \neq (0, 1)$. Then for every $(y, b) \in \mathbb{C} \rtimes \mathbb{T}$ the normal subgroup H of G contains also the element $(y, b)(x, a)(y, b)^{-1} = (y + bx, ba)(-b^{-1}y, b^{-1}) = (y + bx - ay, a)$ and hence contains the coset $\mathbb{C} \times \{a\}$. Being a subgroup, H also contains the set $(\mathbb{C} \times \{a\}) \cdot (\mathbb{C} \times \{a\})^{-1} = \mathbb{C} \times \{1\}$. Taking into account that the quotient group $G/(\mathbb{C} \times \{1\})$ is compact, we conclude that G/H is compact too. Consequently, $h(G) = \tilde{h}(G/H)$ is compact and closed in the Hausdorff space Y, which contradicts our assumption. This contradiction shows that the subgroup H is trivial and the homomorphism h is injective. □

Claim 5. *The map $h : G \to Y$ is a topological embedding.*

Proof. Since h is injective, the family $\tau = \{h^{-1}(U) : U \text{ is open in } Y\}$ is a Hausdorff topology turning G into a paratopological group. We need to show that τ coincides with the standard locally compact topology of the group G. Since the topology τ is weaker than the original product topology of $G = \mathbb{C} \rtimes \mathbb{T}$, the compact set $\{1\} \times \mathbb{T}$ remains compact in the topology τ. Then we can find a neighborhood $U \in \tau$ of the unit $(0, 1) \in G$ such that UU is disjoint with the compact set $\{1\} \times \mathbb{T}$.

Using the compactness of the set $\{0\} \times \mathbb{T} \subset G$ and the continuity of the multiplication in G, find a neighborhood $V \in \tau$ of the unit such that $\{ava^{-1} : v \in V, \ a \in \{0\} \times \mathbb{T}\} \subset U$. We claim that $V \subset \{(z, a) \in \mathbb{C} \rtimes \mathbb{T} : |z| \leq 1\}$. Assuming the opposite, we could find an element $(v, a) \in V$ with $|v| > 1$.

Since $|v^{-1}| < 1$ and $|a| = 1$, there are two complex numbers $b, c \in \mathbb{T}$ such that $b + ac = v^{-1} \in \mathbb{C}$. Observe that $(bv, a) = (0, b)(v, a)(0, b^{-1}) \in U$ and similarly $(cv, a) \in U$. Then

$$UU \ni (bv, a)(cv, a) = (bv + acv, a^2) = (1, a^2) \in \{1\} \times \mathbb{T},$$

which contradicts the choice of the neighborhood U. This contradiction shows that $V \subset \{(z, a) \in G : |z| \leq 1\}$ and hence V has compact closure in the spaces G and (G, τ). This means that the paratopological group (G, τ) is locally compact and, by the Ellis Theorem ([27] 2.3.12), is a topological group. By the Open Mapping Principle 38, the identity homomorphism $G \to (G, \tau)$ is a topological isomorphism and so is the homomorphism $h : G \to h(G) \subset Y$. □

Since the topological group G is Weil-complete (being locally compact), Theorem 17 guarantees that $Y \setminus h(G)$ is an ideal of the semigroup Y. Choose any element $y \in Y \setminus h(G)$ and observe that for the compact subset $K = h(\{0\} \times \mathbb{T}) \subset h(G) \subset Y$ the compact set yKy is contained in the ideal $Y \setminus h(G)$ and hence does not intersect K. By the Hausdorff property of the topological semigroup Y and the compactness of K, the point y has a neighborhood $V \subset Y$ such that $(VKV) \cap K = \varnothing$. Now take any element $v \in V \cap h(G)$ and find an element $(x, a) \in G$ with $h(x, a) = v$. Let $b = -a^{-1} \in \mathbb{T}$ and observe that $(x, a)(0, b)(x, a) = (x, ab)(x, a) = (x, -1)(x, a) = (0, -a)$. Then for the element $k = h(0, b) \in K$ we get $VKV \ni vkv = h(0, -a) \in K$, which contradicts the choice of the neighborhood V. This contradiction completes the proof of the h:TS-closedness of the Lie group $\mathbb{C} \rtimes \mathbb{T}$.

By Theorem 15, the non-compact ω-narrow h:TG-closed group $\mathbb{C} \rtimes \mathbb{T}$ is not MAP. □

The following striking example of Klyachko, Olshanskii and Osin ([29] 2.5) shows that the p:TS-closedness does not imply compactness even for 2-generated discrete topological groups. A group is called 2-*generated* if it is generated by two elements.

Example 4 (Klyachko, Olshanskii, Osin). *There exists a p:TG-closed infinite simple 2-generated discrete topological group G of finite exponent. By Theorems 14 and 32, G is p:TS-closed and is c:TG-closed.*

We do not know if the groups constructed by Klyachko, Olshanskii and Osin can be c:TS-closed. So, we ask

Problem 10. *Is each c:TS-closed topological group compact?*

Finally, we present an example showing that an h:TG-closed topological group needs not be e:TS-closed.

Example 5. *The symmetric group $\mathrm{Sym}(\omega)$ endowed with the topology of pointwise convergence has the following properties:*

(1)   *X is Polish, minimal, and not compact;*
(2)   *X is complete but not Weil-complete;*
(3)   *X is h:TG-closed;*
(4)   *X is not e:TS-closed.*

Proof. The group $\mathrm{Sym}(\omega)$ is Polish, being a G_δ-subset of the Polish space ω^ω. The minimality of the group $\mathrm{Sym}(\omega)$ is a classical result of Gaughan [39]. Being Polish, the topological group $\mathrm{Sym}(\omega)$ is complete. By ([23] 7.1.3), the topological group $\mathrm{Sym}(\omega)$ is not Weil-complete. By Theorem 16, the topological group $\mathrm{Sym}(\omega)$ is not e:TS-closed.

It remains to show that the topological group $X = \mathrm{Sym}(\omega)$ is h:TG-closed. Let $h : X \to Y$ be a continuous homomorphism to a topological group Y. By ([23] 7.1.2), the group $\mathrm{Sym}(\omega)$ is topologically simple, which implies that the kernel $H = h^{-1}(1)$ of the homomorphism h is either trivial or coincides with X. In the second case the group $h(X)$ is trivial and hence closed in Y. In the first case, the homomorphism h is injective. By the minimality of X, the homomorphism h is an isomorphic topological embedding. The completeness of X ensures that the image $h(X)$ is closed in Y. □

Acknowledgments: The author expresses his thanks to Serhiĭ Bardyla and Alex Ravsky for fruitful discussions on the topic of the paper, to Dikran Dikranjan for helpful comments on the initial version of the paper, to Michael Megrelishvili for a valuable remark on the minimality of the Weyl-Heisenberg group, and to the anonymous referee for pointing out the papers [4–6] discussing absolutely closed Lie groups.

Conflicts of Interest: The author declares no conflict of interest.

References

1. Raikov, D.A. On a completion of topological groups. *Izv. Akad. Nauk SSSR* **1946**, *10*, 513–528. (In Russian)
2. Bardyla, S.; Gutik, O.; Ravsky, A. *H*-closed quasitopological groups. *Topol. Appl.* **2017**, *217*, 51–58.
3. Gutik, O.V. On closures in semitopological inverse semigroups with continuous inversion. *Algebra Discret. Math.* **2014**, *18*, 59–85.
4. Omori, H. Homomorphic images of Lie groups. *J. Math. Soc. Jpn.* **1966**, *18*, 97–117.
5. Goto, M. Absolutely closed Lie groups. *Math. Ann.* **1973**, *204*, 337–341.
6. Bader, U.; Gelander, T. Equicontinuous Actions of Semisimple Groups. Preprint. Available online: https://arxiv.org/abs/1408.4217 (accessed on 15 June 2017).
7. Stepp, J.W. A note on maximal locally compact semigroups. *Proc. Am. Math. Soc.* **1969**, *20*, 251–253.
8. Stepp, J.W. Algebraic maximal semilattices. *Pac. J. Math.* **1975**, *58*, 243–248.
9. Dikranjan, D.; Tonolo, A. On a characterization of linear compactness. *Riv. Mat. Pura Appl.* **1995**, *16*, 95–106.
10. Dikranjan, D.; Uspenskij, V.V. Categorically compact topological groups. *J. Pure Appl. Algebra* **1998**, *126*, 149–168.
11. Lukàcs, A. *Compact-Like Topological Groups*; Heldermann Verlag: Lemgo, Germany, 2009.
12. Dikranjan, D.; Shakhmatov, D. Selected topics from the structure theory of topological groups. In *Open Problems in Topology 2*; Pearl, E., Ed.; Elsevier: Amsterdam, The Netherlands, 2007; pp. 389–406.
13. Banakh, T.; Ravsky, A. On *H*-closed paratopological groups. In Proceedings of the Ukrainian Congress of Mathematics, Kyiv, Ukraine, 21–23 August 2001.
14. Ravsky, A. On *H*-closed paratopological groups. *Visnyk Lviv Univ. Ser. Mekh. Mat.* **2003**, *61*, 172–179.
15. Bardyla, S.; Gutik, O. On *H*-complete topological semilattices. *Mat. Stud.* **2012**, *38*, 118–123.
16. Chuchman, I.; Gutik, O. On *H*-closed topological semigroups and semilattices. *Algebra Discret. Math.* **2007**, *1*, 13–23.
17. Gutik, O.; Repovš, D. On linearly ordered *H*-closed topological semilattices. *Semigroup Forum* **2008**, *77*, 474–481.
18. Banakh, T.; Bardyla, S. Completeness and Absolute *H*-Closedness of Topological Semilattices. Preprint. Available online: https://arxiv.org/abs/1702.02791 (accessed on 25 June 2017).
19. Banakh, T.; Bardyla, S. Characterizing Chain-Compact and Chain-Finite Topological Semilattices. Preprint. Available online: https://arxiv.org/abs/1705.03238 (accessed on 25 June 2017).
20. Banakh, T. A Quantitative Generalization of Prodanov-Stoyanov Theorem on Minimal Abelian Topological Groups. Preprint. Available online: https://arxiv.org/abs/1706.05411 (accessed on 20 June 2017).
21. Zelenyuk, E.G.; Protasov, I.V. Complemented topologies on abelian groups. *Sibirsk. Mat. Zh.* **2001**, *42*, 550–560.
22. Prodanov, I.; Stojanov, L.N. Every minimal abelian group is precompact. *C. R. Acad. Bulg. Sci.* **1984**, *37*, 23–26.
23. Dikranjan, D.; Prodanov, I.; Stoyanov, L. Topological Groups: Characters Dualities and Minimal Group Topologies. In *Monographs and Textbooks in Pure and Applied Mathematics*, 2nd ed.; Marcel Dekker: New York, NY, USA, 1989; Volume 130.
24. Dikranjan, D. Recent advances in minimal topological groups. *Topol. Appl.* **1998**, *85*, 53–91.
25. Dikranjan, D.; Megrelishvili, M. Minimality conditions in topological groups. In *Recent Progress in General Topology III*; Hart, K.P., van Mill, J., Simon, P., Eds.; Springer (Atlantis Press): Berlin, Germany, 2014; pp. 229–327.
26. Guran, I.I. Topological groups similar to Lindelöf groups. *Dokl. Akad. Nauk SSSR* **1981**, *256*, 1305–1307.
27. Arhangel'skii, A.; Tkachenko, M. Topological groups and related structures. In *Atlantis Studies in Mathematics*, 1; Atlantis Press: Paris, France; World Scientific Publishing Co. Pte. Ltd.: Hackensack, NJ, USA, 2008.
28. Dikranjan, D.; Prodanov, I. Totally minimal topological groups. *Annuaire Univ. Sofia Fac. Math. Méc.* **1974**, *69*, 5–11.
29. Klyachko, A.; Olshanskii, A.; Osin, D. On topologizable and non-topologizable groups. *Topol. Appl.* **2013**, *160*, 2104–2120.
30. Roelcke, W.; Dierolf, S. *Uniform Structures on Topological Groups and Their Quotients*; McGrawHill: New York, NY, USA, 1981.
31. Engelking, R.; *General Topology*, 2nd ed.; Heldermann: Berlin, Germany, 1989.

32. Stroppel, M. *Locally Compact Groups*; European Mathematical Society: Zürich, Switzerland, 2006.
33. Protasov, I.; Banakh, T. *Ball Structures and Colorings of Graphs and Groups*; VNTL Publishers: Lviv, Ukraine, 2003.
34. Dikranjan, D.; Megrelishvili, M. Relative minimality and co-minimality of subgroups in topological groups. *Topol. Appl.* **2010**, *157*, 62–76.
35. Megrelishvili, M. Generalized Heisenberg groups and Shtern's question. *Georgian Math. J.* **2004**, *11*, 775–782.
36. Rogers, J. *Analytic Sets*; Academic Press: Cambridge, MA, USA, 1980.
37. Robinson, D. *A Course in the Theory of Groups*; Springer: New York, NY, USA, 1996.
38. Morris, S. *Pontryagin Duality and the Structure of Locally Compact Abelian Groups*; Cambridge University Press: New York, NY, USA, 1977.
39. Gaughan, E.D. Topological group structures of infinite symmetric groups. *Proc. Natl. Acad. Sci. USA* **1967**, *58*, 907–910.

axioms

MDPI

Review

Selective Survey on Spaces of Closed Subgroups of Topological Groups

Igor V. Protasov

Faculty of Computer Science and Cybernetics, Kyiv University, Academic Glushkov pr. 4d, 03680 Kyiv, Ukraine; i.v.protasov@gmail.com

Received: 8 October 2018 ; Accepted: 24 October 2018 ; Published: 26 October 2018

Abstract: We survey different topologizations of the set $\mathcal{S}(G)$ of closed subgroups of a topological group G and demonstrate some applications using Topological Groups, Model Theory, Geometric Group Theory, and Topological Dynamics.

Keywords: space of closed subgroups; Chabauty topology; Vietoris topology; Bourbaki uniformity

MSC: 22A05; 22B05; 54B20; 54D30

For a topological group G, $\mathcal{S}(G)$ denotes the set of all closed subgroups of G. There are many ways to endow $\mathcal{S}(G)$ with a topology related to the topology of G. Among these methods, the most intensively studied are the Chabauty topology, rooted in *Geometry of Numbers*, and the Vietoris topology, based on *General Topology*; both coincide if G is compact. The spaces of closed subgroups are interesting by their own merits, but they also have some deep applications in *Topological Groups and Model Theory, Geometric Group Theory, and Dynamical Systems*. This survey is my subjective look at this area.

1. Chabauty Spaces

1.1. From Minkowski to Chabauty

We recall that a *lattice* L in $\mathbb{R}^n$ is a discrete subgroup of rank n. We define *min L* as the length of the shortest nonzero vector from L, and we define *vol* $(\mathbb{R}^n/L)$ as the volume of a basic parallelepiped of L.

A sequence $(L_m)_{m \in \omega}$ of lattices in $\mathbb{R}^n$ converges to the lattice L if, for each $m \in \omega$, one can choose a basis $a_1(m), \ldots, a_n(m)$ of L_m and a basis $a_1, \ldots, a_n$ of L such that the sequence $(a_i(m))_{m \in \omega}$ converges to a_i for each $i \in \{1, \ldots, n\}$. This convergence of lattices was introduced by H. Minkowski [1], and its usage in *Geometry of Numbers* (see [2]) is based on the following theorem from K. Mahler [3].

Theorem 1. *Let $\mathcal{M}$ be a set of lattices in $\mathbb{R}^n$. Every sequence in $\mathcal{M}$ has a convergent subsequence if and only if there exist two constants, $C > 0$ and $c > 0$, such that min $L > c$, vol $(\mathbb{R}^n \setminus L) < C$ for each $L \in \mathcal{M}$.*

What we know now is that Chabauty topology was invented by C. Chabauty in [4] in order to extend Theorem 1 to lattices in connected Lie groups. A discrete subgroup L of a connected Lie group G is called a *lattice* if the quotient space G/L is compact.

Let X be a Hausdorff locally compact space, and let $exp\ X$ denote the set of all closed subsets of X. The sets

$$\{F \in exp\ X : F \cap K = \varnothing\},\ \{F \in exp\ X : F \cap U \neq \varnothing\},$$

where *K* runs over all compact subsets of *X* and *U* runs over all open subsets of *X*, form the subbase of the *Chabauty topology* on $exp\ X$. The space $exp\ X$ is compact and Hausdorff. If *X* is discrete, then $exp\ X$ is homeomorphic to the Cantor cube $\{0,1\}^{|X|}$.

We note also that a net $(F_\alpha)_{\alpha \in \mathcal{I}}$ converges in $exp\ X$ to *F* if and only if

- for every compact *K* of *X* such that $K \cap F = \varnothing$, there exists $\beta \in \mathcal{I}$ such that $F_\alpha \cap K = \varnothing$ for each $\alpha > \beta$;

- for every $x \in F$ and every neighborhood *U* of *x*, there exists $\gamma \in \mathcal{I}$ such that $F_\alpha \cap U \neq \varnothing$ for each $\alpha > \gamma$.

If *G* is a locally compact group, then $\mathcal{S}(G)$ is a closed subspace of $exp\ G$ (so, $\mathcal{S}(G)$ is compact); $\mathcal{S}(G)$ is called the *Chabauty space* of *G*.

Theorem 2. *[4]. Let G be a connected unimodular Lie group. A set $\mathcal{M}$ of lattices in G is relatively compact in $\mathcal{M}$ if and only if there exists a constant $C > 0$ and a neighborhood U with the identity e of G such that $L \cap U = \{e\}$ and vol $(G/L) < C$ for each $L \in \mathcal{M}$.*

There was some technical improvement made in [5] and the paper [4], which is included in [6], Chapter 8.

1.2. Pontryagin–Chabauty Duality

This duality was established in [7] and detailed in [8]. We use the standard abbreviation LCA for a locally compact Abelian group. Let *G* be an LCA-group $G^\wedge$, let denote its dual group $G^\wedge = Hom\ (G, \mathbb{R}/\mathbb{Z})$, and let φ denote the canonical bijection $\mathcal{S}(G) \longrightarrow \mathcal{S}(G^\wedge)$, $\varphi(X) = \{f \in G^\wedge : X \subseteq ker\ f\}$.

Theorem 3. *For every LCA-group G, the bijection $\varphi : \mathcal{S}(G) \longrightarrow \mathcal{S}(G^\wedge)$ is a homeomorphism.*

Typically, Theorem 3 is applied to replace $\mathcal{S}(G)$ by $\mathcal{S}(G^\wedge)$ in the case of a compact Abelian group *G*. In what follows, we use the following notations: $\mathbb{C}_n$ is a cyclic group of order *n*, $\mathbb{C}_{p^\infty}$ is a quasi-cyclic (or Prüfer) *p*-group, $\mathbb{Z}$ is a discrete group of integers, $\mathbb{Z}_p$ is the group of *p*-adic integers, and $\mathbb{Q}_p$ is an additive group of a field of *p*-adic numbers.

1.3. S(G) for Compact G

The following two lemmas from [9] are the basic technical tools in this area.

Lemma 1. *If G, H are compact groups and $\varphi : G \longrightarrow H$ is a continuous surjective homomorphism, then the mapping $\mathcal{S}(G) \longrightarrow \mathcal{S}(H)$, $X \longmapsto \varphi(X)$ is continuous and open.*

The continuity is easily deduced, but to prove the openness, we need

Lemma 2. *Let G be a compact group, $X \in \mathcal{S}(G)$. Then, the following subsets form a base of the neighborhoods of X is $\mathcal{S}(G)$:*

$$\mathcal{N}_X(U, N, x_1, \ldots, x_n) = \{u^{-1}Yu : u \in U,\ Y \in \mathcal{S}(G),\ Y \subseteq XN,\ Y \cap x_1 U \neq \varnothing, \ldots, Y \cap x_n U \neq \varnothing,\}$$

where U is a neighborhood of the identity of G, N is closed normal subgroup such that G/N is a Lie group, and $x_1, \ldots, x_n$ are arbitrary elements of X, $n \in \mathbb{N}$.

In particular, if G is a compact Lie group, then Lemma 2 states that there is a neighborhood $\mathcal{N}$ of X such that each subgroup $Y \in \mathcal{N}$ is conjugated to some subgroup of X. The Montgomery–Yang theorem on tubes [10] (see also ([11], Theorem 5.4, Chapter 2)) plays the key role in the proof of Lemma 2.

We recall that the *cellularity* (or Souslin number) $c(X)$ of a topological space X is the supremum of cardinalities of disjoint families of open subsets of X. A topological space X is called *dyadic* if X is a continuous image of some Cantor cube $\{0,1\}^\kappa$.

The *weight* $w(X)$ of a topological space X is the minimal cardinality of the open bases of X.

Theorem 4. *[9]. For every compact group G, we have $c(\mathcal{S}(G)) \leq \aleph_0$. In addition, if $w(G) \leq \aleph_1$, then $\mathcal{S}(G)$ is dyadic.*

Theorem 5. *[12]. Let a group G be either profinite or compact and Abelian. If $w(G) > \aleph_2$, then the space $\mathcal{S}(G)$ is not dyadic.*

Theorem 6. *[12]. Let G be an infinite compact Abelian group such that $w(G) \leq \aleph_1$. Then, the space $\mathcal{S}(G)$ is homeomorphic to the Cantor cube $\{0,1\}^{w(G)}$ if and only if $\mathcal{S}(G)$ has no isolated points.*

An Abelian group G is called *Artinian* if every increasing chain of subgroups of G is finite; every such group is isomorphic to the direct *sum* $\oplus_{p \in F} \mathbb{C}_{p^\infty} \oplus K$, where F is a finite set of primes, and K is a finite subgroup. An Abelian group G is called *minimax* if G has a finitely generated subgroup N such that G/N is Artinian.

Theorem 7. *[12]. For a compact Abelian group G, the space $\mathcal{S}(G)$ has an isolated point if and only if the dual group $G^\wedge$ is minimax.*

1.4. $\mathcal{S}(G)$ for LCA G

The space $\mathcal{S}(\mathbb{R})$ is homeomorphic to the segment $[0,1]$. By [13], $\mathcal{S}(\mathbb{R}^2)$ is homeomorphic to the sphere $\mathbf{S}^4$. For $n \geq 3$, $\mathcal{S}(\mathbb{R}^n)$ is not a topological manifold and its structure is far from understood (see [14]).

Theorem 8. *[15]. The space $\mathcal{S}(G)$ of an LCA-group G is connected if and only if G has a subgroup topologically isomorphic to $\mathbb{R}$.*

If F is a non-solvable finite group, then $\mathcal{S}(\mathbb{R} \times F)$ is not connected ([8], Proposition 8.6).

Theorem 9. *[8]. The space $\mathcal{S}(G)$ of an LCA-group G is totally disconnected if and only if G is either totally disconnected or each element of G belongs to a compact subgroup.*

Some more information on $\mathcal{S}(G)$ for LCA G can be found in [8] and the references therein, particularly on the topological dimension of $\mathcal{S}(G)$.

By Theorems 3 and 4, $c(\mathcal{S}(G)) \leq \aleph_0$ for every discrete Abelian group. We say that a topological space X has the *Shanin number ω* if any uncountable family $\mathcal{F}$ of the non-empty open subsets of X has an uncountable subfamily $\mathcal{F}'$ such that $\cap \mathcal{F}' \neq \emptyset$. Evidently, if a space X has the Shanin number ω, then $c(X) \leq \aleph_0$. By [16], Theorem 1, for every discrete Abelian group G, the space $\mathcal{S}(G)$ has the Shanin number ω. By [16], Theorem 3, for every infinite cardinal τ, there exists a solvable discrete group G such that $c(\mathcal{S}(G)) = |G| = \tau$.

1.5. S(G) as a Lattice

The set $S(G)$ has the natural structure of a lattice with the operations $\vee$ and $\wedge$, where $A \wedge B = A \cap B$, and $A \vee B$ is the smallest closed subgroup of G containing A and B. In this subsection, we formulate some results from [17] about the interrelations between the topological and lattice structures on $S(G)$.

For $g \in G$, $\overline{< g >}$ denotes the subgroup of G topologically generated by g. A totally disconnected locally compact group G is called *periodic* if $\overline{< g >}$ is compact for each $g \in G$. In this case, $\pi(G)$ denotes the set of all prime numbers such that $p \in \pi(G)$ if and only if $g \in G$ such that $\overline{< g >}$ is topologically isomorphic either to $\mathbb{C}_{p^n}$ or to $\mathbb{Z}_p$; this g is called a *topological p-element*.

Theorem 10. *For a compact group G, the following statements are equivalent:*

(i) $\wedge$ *is continuous;*

(ii) $\wedge$ *and $\vee$ are continuous;*

(iii) *G is the semidirect product $K \rtimes P$, where K is profinite with finite Sylow p-subgroups, P is Abelian profinite and each Sylow p-subgroup of G is $\mathbb{Z}_p$, $\pi(K) \cap \pi(P) = \varnothing$, and the centralizer of each Sylow p-subgroup of G has a finite index in G.*

Theorem 11. *For a locally compact group G, the operation $\wedge$ is continuous if and only if the following conditions are satisfied:*

(i)   *G is either discrete or periodic;*

(ii) $\wedge$ *is continuous in $\mathcal{S}(H)$ for each compact subgroup H of G;*

(iii) *the centralizer of each topological p-element of G is open.*

We recall that a torsion group G is *layer-finite* if the set $\{g \in G : g^n = e\}$ is finite for each $n \in \mathbb{N}$. A layer-finite group G is called *thin* if each Sylow p-subgroup of G is finite (equivalently, G has no subgroup isomorphic to $\mathbb{C}_{p^\infty}$).

Theorem 12. *Let G be a locally compact group. The operations $\wedge$ and $\vee$ are continuous if and only if G is periodic and topologically isomorphic to $A \times B \times (C \rtimes D)$, where C has a dense thin layer-finite subgroup; A, B, D are Abelian with Sylow p-subgroups $\mathbb{C}_{p^\infty}$, $\mathbb{Q}_p$, or $\mathbb{Z}_p$; the sets $\pi(A)$, $\pi(B)$, $\pi(G)$, $\pi(D)$ are pairwise disjoint; and the centralizer of each Sylow p-subgroup of G is open.*

1.6. From Chabauty to Local Method

A topological group G is called *topologically simple* if each closed normal subgroup of G is either G or $\{e\}$. Every topologically simple LCA-group is discrete, and either $G = \{e\}$ or G is isomorphic to $\mathbb{C}_p$.

Following the algebraic tradition, we say that a group G is *locally nilpotent (solvable)* if every finitely generated subgroup is nilpotent (solvable).

In [18], Problem 1.76, V. Platonov asked whether there exists a non-Abelian, topologically simple, locally compact, locally nilpotent group. Here, we present the negative answer to this question for the locally solvable group obtained in [19].

Let G be a locally compact, locally solvable group. We take $g \in G \setminus \{e\}$, choose a compact neighborhood U of G, and denote by $\mathcal{F}$ the family of all topologically finitely generated subgroups of G containing g. We may assume that G is not topologically finitely generated, so $\mathcal{F}$ is directed by the inclusion $\subset$. For each $F \in \mathcal{F}$, we choose $A_F, B_F \in \mathcal{S}(F)$ such that $B_F \subset A_F$; A_F and B_F are normal in F, $A_F \cap U \neq \varnothing$, $B_F \cap U = \varnothing$, and A_F / B_F is Abelian. Since $\mathcal{S}(G)$ is compact, we can choose two subnets $(A_\alpha)_{\alpha \in \mathcal{I}}$, $(B_\alpha)_{\alpha \in \mathcal{I}}$ of the nets $(A_F)_{F \in \mathcal{F}}$, $(B_F)_{F \in \mathcal{F}}$ which converge to $A, B \in \mathcal{S}(G)$. Then A, B are normal in G, and A/B is Abelian. Moreover, $x \notin B$ and $A \cap U \neq \emptyset$. If $A \neq \{G\}$, then A is a proper normal subgroup of G; otherwise, G/B is Abelian.

In [20], the Chabauty topology was defined on some systems of closed subgroups of a locally compact group G. A system $\mathfrak{A}$ of closed subgroups of G is called *subnormal* if

- $\mathfrak{A}$ contains $\{e\}$ and G;
- $\mathfrak{A}$ is linearly ordered by the inclusion $\subset$;
- for any subset $\mathfrak{M}$ of $\mathfrak{A}$, the closure of $\bigcup_{F \in \mathfrak{M}} F \in \mathfrak{A}$ and $\bigcap_{F \in \mathfrak{M}} F \in \mathfrak{A}$;
- whenever A and B comprise a jump in $\mathfrak{A}$ (i.e., $B \subset A$, and no members of $\mathfrak{A}$ lie between B and A), B is a normal subgroup of A.

If the subgroups A, B form a jump, then A/B is called a factor of G. The system is called *normal* if each $A \in \mathfrak{A}$ is normal in G.

A group G is called an RN-group if G has a normal system with Abelian factors. Among the local theorems from [20], one can find the following: if every topologically finitely generated subgroup of a locally compact group G is an RN-group, then G is an RN-group. In particular, every locally compact, locally solvable group is an RN-group.

In 1941 (see ([21], pp. 78–83), A.I. Mal'tsev obtained local theorems for discrete groups as applications of the following general local theorem: if every finitely generated subsystem of an algebraic system A satisfies some property $\mathcal{P}$, which can be defined by some quasi-universal second-order formula, then A satisfies $\mathcal{P}$.

In [22], Mal'tsev's local theorem was generalized on a topological algebraic system. The part of the model-theoretical Compactness Theorem in Mal'tsev's arguments employs some convergents of closed subsets. A net $(F_\alpha)_{\alpha \in \mathcal{I}}$ of closed subsets of a topological space X S-converges to a closed subset F if

- for every $x \in F$ and every neighborhood U of x, there exists $\beta \in \mathcal{I}$ such that $F_\alpha \cap U \neq \emptyset$ for each $\alpha > \beta$;
- for every $y \in X \setminus F$, there exist a neighborhood V of y and a $\gamma \in \mathcal{I}$ such that $F_\alpha \cap V = \emptyset$ for each $\alpha > \gamma$.

Every net of closed subsets of an arbitrary (!) topological space has a convergent subnet. If X is a Hausdorff locally compact space, then the S-convergence coincides with the convergence in the Chabauty topology.

1.7. Spaces of Marked Groups

Let F_k be the free group of rank k, with the free generators $x_1, \ldots, x_k$, and let $\mathcal{G}_k$ denote the set of all normal subgroups of F_k. In the metric form, the Chabauty topology on $\mathcal{G}_k$ was introduced in [23] as a reply to Gromov's idea of the topologizations of some sets of groups [24].

Let G be a group generated by $g_1, \ldots, g_k$. The bijection $x_i \longmapsto g_i$ $g_1, \ldots, g_n$ can be extended to the homomorphism $f : F_k \longrightarrow G$. With the correspondence $G \longmapsto \ker f$, $\mathcal{G}_k$ is called the *space marked k-generated groups*.

A couple of papers in development by [23] are aimed at understanding how large, in the topological sense, are the well-known classes of finitely generated groups, or how a given k-generated group is placed in $\mathcal{G}_k$ (see [25]). Among the applications of $\mathcal{G}_k$, we mention the construction of topologizable Tarski Monsters in [26].

1.8. Dynamical Development

Every locally compact group G acts on the Chabauty space $\mathcal{S}(G)$ by the rule: $(g, H) \longmapsto g^{-1}Hg$. Under this action, every minimal closed invariant subset of $\mathcal{S}(G)$ is called a *uniformly recurrent subgroup* (URS). The study of URSs was initiated by Glasner and Weiss [27] with the following observation.

Let G be a locally compact group G acting on a compact X so that is G minimal, i.e., the orbit of each point $x \in X$ is dense. We consider the mapping $Stab : X \longrightarrow \mathcal{S}(G)$, defined by $Stab(x) = \{g \in G : gx = x\}$. Then, there is the unique URS contained in the closure of $Stab(X)$. This URS is called the *stabilizer URS*. Glasner and Weiss asked whether every URS of a locally compact group G arises as the stabilizer URS of a minimal action of G on a compact space. This question was answered in the affirmative in [28].

2. Vietoris Spaces

For a topological space X, the Vietoris topology on the set $exp\, X$ of all closed subsets of X is defined by the subbase of the open sets

$$\{F \in exp\, X : F \subseteq U\}, \ \{F \in exp\, X : F \cap V \neq \varnothing\},$$

where U, V run over all open subsets of X.

A net $(F_\alpha)_{\alpha \in \mathcal{I}}$ converges to F in $exp\, X$ if and only if

- for each open subset U of X such that $F \subseteq U$, there exists $\beta \in \mathcal{I}$ such that $F_\alpha \subseteq U$ for each $\alpha > \beta$;
- for each $x \in F$ and each neighborhood V of x, there exists $\gamma \in \mathcal{I}$ such that $F_\alpha \cap V \neq \varnothing$ for each $\alpha > \gamma$.

If X is regular, then $\mathcal{S}(G)$ is closed in $exp\, G$. To my knowledge, the spaces $\mathcal{S}(G)$, where G needs not be compact, endowed with Vietoris topologies appeared in [29] with the characterization of LCA-groups G such that the canonical mapping $\varphi : \mathcal{S}(G) \longrightarrow \mathcal{S}(G^\wedge)$ is a homeomorphism.

2.1. Compactness

We cannot ask for a constructive description of arbitrary topological groups G with compact space $\mathcal{S}(G)$ because we know nothing about G with $S(G) = 2$.

Theorem 13. *[30]. Let G be a locally compact group. The space $\mathcal{S}(G)$ is compact if and only if G is one of the following groups:*

(i) *G is compact;*
(ii) *$\mathbb{C}_{p_1^\infty} \times \ldots \times \mathbb{C}_{p_n^\infty} \times K$, where $p_1, \ldots, p_n$ are distinct prime numbers, K is finite, and each p_i is not a divisor of $|K|$;*
(iii) *$Q_p \times K$, where K is finite and p does not divide $|K|$.*

A similar characterization of groups with compact $\mathcal{S}(G)$ is given in [31], provided that G has a base of neighborhoods at the identity consisting of subgroups.

Theorem 14. *[32]. Let G be a locally compact group. A closed subset $\mathcal{F}$ of $\mathcal{S}(G)$ is compact if and only if the following conditions are satisfied:*

(i) *every descending chain of non-compact subgroups from F is finite;*

(ii)   *every closed subset $\mathcal{F}'$ of $\mathcal{F}$ has only a finite number of non-compact subgroups maximal in $\mathcal{F}$;*

(iii) *if a closed subset $\mathcal{F}'$ of $\mathcal{F}$ has no non-compact subgroups, then $\cup \mathcal{F}'$ is compact.*

Two corollaries: Every compact subset of $\mathcal{S}(G)$ consisting of non-compact subgroups is scattered; a subset $\mathcal{F}$ is compact if and only if $\mathcal{F}$ is countably compact.

For locally compact groups with the σ-compact space $\mathcal{S}(G)$ (see [33]), a description of the LCA-groups with locally compact space $\mathcal{S}(G)$ can be obtained in [34].

A topological group G is called *inductively compact* if every finite subset of G is contained in a compact subgroup. For a group G, $K(G)$ and $IK(G)$ denote the sets of all compact and closed inductively compact subgroups.

Theorem 15. *[35]. For every locally compact group G, $IK(G)$ is the closure of $K(G)$.*

Two corollaries: If G is a connected Lie group, then $K(G)$ is closed; $\mathcal{S}(G)$ is a k-space for each locally compact group G of countable weight, i.e., the topology of $\mathcal{S}(G)$ is uniquely determined by the family of all compact subsets of $\mathcal{S}(G)$.

2.2. Metrizability and Normality

LCA-groups G with metrizable and normal space $\mathcal{S}(G)$ were characterized by S. Panasyuk in the candidate thesis *Normality and metrizability of the space of closed subgroups*, Kyiv University, 1989.

Theorem 16. *For a discrete Abelian group G, the following statements are equivalent:*

(i)   *$\mathcal{S}(G)$ is metrizable;*

(ii) *$\mathcal{S}(G)$ is normal;*

(iii) *G has a finitely generated subgroup H such that $G/H = \mathbb{C}_{p_1^\infty} \times \ldots \times \mathbb{C}_{p_n^\infty}$, where $p_1, \ldots, p_n$ are distinct primes.*

In the general case, metrizability and normality of $\mathcal{S}(G)$ are not equivalent, but if G is a connected semisimple Lie group, then $\mathcal{S}(G)$ is metrizable if and only if $\mathcal{S}(G)$ is normal if and only if G is compact (see [36,37]). The space $\mathcal{S}(G)$ for every connected solvable Lie group is metrizable [36].

2.3. Some Cardinal Invariants

We remind the reader that $c(X)$ denotes the cellularity of X.

Theorem 17. *[9]. For every infinite locally compact group G, we have $c(\mathcal{S}(G)) \leq c(G)$.*

Theorem 18. *[38]. For every locally compact group G, the following conditions are equivalent:*

(i)   *$\mathcal{S}(G)$ is of countable pseudocharacter;*

(ii) *$\mathcal{S}(G)$ is of countable tightness;*

(iii) *$\mathcal{S}(G)$ is sequential;*

(iv) *$w(G) \leq \aleph_0$.*

3. Other Topologizations

3.1. Bourbaki Uniformities

Let $(X, \mathcal{U})$ be a uniform space. The uniformity $\mathcal{U}$ induces the uniformity $\tilde{\mathcal{U}}$ on the set $\mathcal{F}(X)$ of all non-empty closed subsets of X which have as a base the family of sets of the form

$$\{(A,B) \in \mathcal{F}(X) \times \mathcal{F}(X) : B \subseteq U(A),\ A \subseteq U(B)\},$$

whenever $U \in \mathcal{U}$. The uniformity $\tilde{\mathcal{U}}$ was introduced in [39] (Chapter 2, § 1), and $\tilde{\mathcal{U}}$ is called *the Bourbaki* (sometimes, Hausdorff–Bourbaki) *uniformity*.

Let G be a topological group. We endow G with the left uniformity L and $F(G)$ with the Bourbaki uniformity $\tilde{L}$. We denote by $\mathcal{L}(G)$ and $\mathcal{B}(G)$ the subspaces of $\mathcal{F}(G)$ consisting of all subgroups and all totally bounded subsets of G.

Theorem 19. *[40]. Let G be a group with a base at the identity consisting of subgroups. The space $\mathcal{L}(G)$ is compact if and only if G is totally bounded and $K \cap G$ is dense in K for each closed subgroup K from the completion of G.*

In particular, if $\mathcal{L}(G)$ is compact, then G is totally minimal.

Theorem 20. *[40]. If a group G is complete in the left uniformity, then $\mathcal{B}(G)$ is complete.*

We recall that a topological group G is *almost metrizable* if each neighborhood of e contains a compact subgroup K such that the set of all open subsets containing K have a countable base. Every metrizable and every locally compact topological group is almost metrizable.

Theorem 21. *[40]. If an almost metrizable group G is complete in the left uniformity, then $\mathcal{F}(G)$ is complete.*

In [41], Theorem 21 is proved with the bilateral uniformity on G (and so on $\mathcal{F}(G)$) in place of the left uniformity.

3.2. Functionally Balanced Groups

For a topological group G, the set $\mathcal{F}(G)$ has the natural structure of a semigroup with the operation $(A,B) \longmapsto cl\ AB$.

Theorem 22. *[42]. For a topological group G, the following statements are equivalent:*

(i)   *$\mathcal{F}(G)$ is a topological semigroup;*

(ii) *for every subset X of G and every neighborhood U of e, there exists a neighborhood V of e such that $VX \subseteq XU$;*

(iii) *every bounded left uniformly continuous function on G is right uniformly continuous.*

A topological group G is called *balanced* (or a SIN-group) if the left and right uniformities of G coincide. A group G is called *functionally balanced* if G satisfies (iii) of Theorem 22. The study of functionally balanced groups was initiated by G. Itzkowitz [43].

The equivalence of (ii) and (iii) in Theorem 22 is a criterion for a topological group to be functionally balanced. In [44], this criterion was used to show that each almost-metrizable functionally balanced group is balanced.

3.3. Lattice Topologies

These topologies on a complete lattice $\mathcal{L}(G)$ of closed subgroups are algebraically defined by the lattice structure of $\mathcal{L}(G)$.

For example, a net $(A_\alpha)_{\alpha \in \mathcal{I}}$ in $\mathcal{L}(G)$ *order-converges* to $A \in \mathcal{L}(G)$ if there exist two nets $(B_\alpha)_{\alpha \in \mathcal{I}}$, $(C_\alpha)_{\alpha \in \mathcal{I}}$ in $\mathcal{L}(G)$ such that, for each $\alpha \in \mathcal{I}$, $B_\alpha \subseteq A_\alpha \subseteq C_\alpha$ and $\vee_{\alpha \in \mathcal{I}} B_\alpha = \wedge_{\alpha \in \mathcal{I}} C_\alpha = A$. By [45], for a compact group G, every net in $\mathcal{L}(G)$ has an order-convergent subset if and only if $\mathcal{L}(G)$ endowed with the Shabauty topology is a topological lattice (see Theorem 10).

More on the lattices' topologies on $\mathcal{L}(G)$ in the case of a compact G can be found in [46].

3.4. Segment Topologies

Let G be a topological group; $\mathcal{P}_G$ is the family of all subsets of G, and $[G]^{<\omega}$ is the family of all finite subsets of G. Each pair $\mathcal{A}, \mathcal{B}$ of subsets of $\mathcal{P}_G$ closed under finite unions defines the segment topology on $\mathcal{L}(G)$ with a base consisting of the segments

$$[A, G \setminus B] = \{X \in \mathcal{L}(G) : A \subseteq X \subseteq G \setminus B\}, \ A \in \mathcal{A}, \ B \in \mathcal{B}.$$

These topologies are described in [47] in the following three cases: $\mathcal{A} = \mathcal{B} = [G]^{<\omega}$; $\mathcal{A} = \mathcal{P}_G$ and $\mathcal{B} = [G]^{<\omega}$; $\mathcal{A} = [G]^{<\omega}$, $\mathcal{B} = \mathcal{P}_G$

3.5. (Σ, Θ)-Topologies

This general construction for topologizations of the set $\mathcal{L}(G)$ of closed subgroups of a topological group G from [48] produces Chabauty, Vietoris, and Bourbaki topologies, along with plenty of other topologies.

We assume that, for each $H \in \mathcal{L}(G)$, $\Sigma(H)$ is some family of open subsets of G, $\Sigma = \cup_{H \in \mathcal{L}(G)} \Sigma(H)$, and the following conditions are satisfied:

- if $U, V \in \Sigma(H)$, then $U \cap V$ contains some $W \in \Sigma(H)$;
- for every $U \in \Sigma(H)$, there exists $V \in \Sigma(H)$ such that $U \in \Sigma(K)$ for each $K \in \mathcal{L}(G)$, $K \subseteq V$;
- $\cap_{U \in \Sigma(H)} \overline{U} = H$ for each $H \in \mathcal{L}(G)$.

Then, the family $\{X \in \mathcal{L}(G) : X \subseteq U\}$, $U \in \Sigma$, is a base for the Σ-*topology* on $\mathcal{L}(G)$.

Let τ denote the topology of G, and let $\mathcal{P}_\tau$ denote the family of all subsets of τ. We assume that, for each $H \in \mathcal{L}(G)$, $\Theta(H)$ is some subset of $\mathcal{P}_\tau$ such that the following conditions are satisfied:

- for every $\alpha, \beta \in \Theta(H)$, there is a $\gamma \in \Theta(H)$ such that $\alpha < \gamma$, $\beta < \gamma$ ($\alpha < \beta$ means that, for every $U \in \alpha$, there exists $V \in \beta$ such that $V \subseteq U$);
- for every $\alpha \in \Theta(H)$, there exists $\beta \in \Theta(H)$ such that if $K \in \mathcal{L}(G)$ and $K \cap V \neq \varnothing$ for each $V \in \beta$, then $\alpha < \gamma$ for some $\gamma \in \Theta(K)$;
- for each $H \in \mathcal{L}(G)$ and every neighborhood V of x, there exists $\alpha \in \Theta(H)$ such that $x \in U$, $U \subseteq V$ for some $U \in \alpha$.

Then, the family $\{X \in \mathcal{L}(G) : X \cap U \neq \varnothing$ for each $U \in \alpha\}$, where $\alpha \in \Theta(H)$, $H \in \mathcal{L}(G)$, is a base for the Θ-*topology* on $\mathcal{L}(G)$.

The upper bound of Σ- and Θ-topologies is called the (Σ, Θ)-*topology*.

A net $(H_\alpha)_{\alpha \in \mathcal{I}}$ converges in (Σ, Θ)-topology to $H \in \mathcal{L}(G)$ if and only if

- for any $U \in \Sigma(H)$, there exists $\beta \in \mathcal{I}$ such that $H_\alpha \subseteq U$ for each $\alpha > \beta$;
- for any $\alpha \in \Theta(H)$, there exists $\gamma \in \mathcal{I}$ such that $H_\alpha \cap V \neq \varnothing$ for each $\alpha > \gamma$.

In [48], one can find characterizations of G with a compact and discrete $\mathcal{L}(G)$ for some concrete (Σ, Θ)-topologies.

3.6. Hyperballeans of Groups

Let G be a discrete group. The set $\{Fg : g \in G, F \in [G]^{<\omega}\}$ is a family of balls in the finitary coarse structure on G. For definitions of coarse structures and balleans, see [49,50]. The finitary coarse structure on G induces the coarse structure on $\mathcal{L}(G)$ in which $\{X \in \mathcal{L}(G) : X \subseteq FA, A \in FX\}, F \in [G]^{<\omega}$ is the family of balls centered at $A \in \mathcal{L}(G)$. The set $\mathcal{L}(G)$ endowed with the finitary coarse structure is called a hyperballean of G. Hyperballeans of groups, carefully studied in [51], can be considered as asymptotic counterparts of Bourbaki uniformities.

Funding: This research received no external funding.

Conflicts of Interest: The author declares no conflicts of interest.

References

1. Minkowski, H. *Geometrie der Zahlen*; R.G. Teubner: Leipzig/Berlin, Germany, 1910.
2. Cassels, J.W.S. *An Introduction to the Geometry of Nombres*; Classics in Mathematics; Springer: Berlin, Germany, 1997.
3. Mahler, K. On lattice points in n-dimensional star bodies: I. Existence theorems. *Proc. R. Soc. Lond.* **1946**, *187*, 151–187. [CrossRef]
4. Chabauty, C. Limite d'ensembles et geometrie des nombres. *Bull. Soc. Math. Fr.* **1950**, *78*, 143–151. [CrossRef]
5. Macbeath, A.M.; Swierczkowski, S. Limits of lattices in a compactly generated group. *Can. J. Math.* **1960**, *12*, 427–437. [CrossRef]
6. Bourbaki, N. *Éléments de Mathématique. Fascicule XXIX. Livre VI: Intégration. Chapitre 7: Measure de Haar. Chapitre 8: Convolution et Représentations*; Actualites Scientifiques et Industrielles 1306; Hermann: Paris, France, 1963.
7. Protasov, I.V. Dualisms of topological abelian groups. *Ukr. Math. J.* **1979**, *31*, 207–211. [CrossRef]
8. Cornulier, Y. On the Chabauty space of locally compact abelian group. *Algebr. Geom. Topol.* **2011**, *11*, 2007–2035. [CrossRef]
9. Protasov, I.V. On the Souslin number of the space of subgroups of a locally compact group. *Ukr. Math. J.* **1988**, *40*, 654–658. [CrossRef]
10. Montgomery, D.; Yang, C.T. The existence of a slice. *Ann. Math.* **1957**, *65*, 108–116. [CrossRef]
11. Bredon, G.E. *Introduction to Compact Transformation Groups*; Academic Press: New York, NY, USA; London, UK, 1972.
12. Tsybenko, Y.V. Dyadic spaces of subgroups of a topological group.*Ukr. Math. J.* **1986**, *38*, 635–639.
13. Pourezza, I.; Hubbard, J. The space of closed subgroup of $\mathbb{R}^2$. *Topology* **1979**, *18*, 143–146. [CrossRef]
14. Kloeckner, B. The space of closed subgroups of $\mathbb{R}^n$ is stratified and simply connected. *J. Topol.* **2009**, *2*, 570–588. [CrossRef]
15. Protasov, I.V.; Tsybenko, Y.V. Connectedness in the space of subgroups. *Ukr. Math. J.* **1983**, *35*, 382–385. [CrossRef]
16. Leiderman, A.; Protasov, I.V. Cellularity of a space of subgroups of a discrete group. *Comment. Math. Univ. C* **2008**, *49*, 519–522.
17. Protasov, I.V.; Tsybenko, Y.V. Chabauty topology in the lattice of closed subgroups. *Ukr. Math. J.* **1984**, *36*, 207–213. [CrossRef]
18. Mazurov, V.D.; Khukhro, E.I. (Eds.) *Unsolved Problems in Group Theory*; 13th ed.; Russian Academy of Sciences Siberian Division, Institute of Mathematics: Novosibirsk, Russia, 1995.
19. Protasov, I.V. Closed invariant subgroups of locally compact groups. *Dokl. Acad. Nauk SSSR* **1978**, *239*, 1060–1062.
20. Protasov, I.V. Local theorems for topological groups. *Izv. Acad. Nauk SSSR. Ser. Mat.* **1979**, *43*, 1430–1440. [CrossRef]

21. Mal'tsev, A.I. *Selected Works. Classic Algebra*; Nauka: Moscow, Russia, 1976; Volume 1.
22. Protasov, I.V. Local method and compactness theorem for topological algebraic systems. *Sib. Math. J.* **1982**, *23*, 136–143. [CrossRef]
23. Grigorchuk, R.I. Degrees of growth of finitely generated groups and the theory of invariant means. *Math. USSR Izv.* **1985**, *25*, 259–330. [CrossRef]
24. Gromov, M. Groups of polynomial growth and expanding maps. *Inst. Hautes Etudes Sci. Publ. Math.* **1981**, *53*, 53–73. [CrossRef]
25. Cornulier, Y.; Guyot, L.; Pitsch, W. On the isolated points in the space of groups. *J. Algebra* **2007**, *307*, 254–277. [CrossRef]
26. Klyachko, A.A.; Olshanskii, A.Y.; Osin, D.V. On topologizable and non topologizable groups. *Topol. Appl.* **2013**, *160*, 2014–2020. [CrossRef]
27. Glasner, E.; Weiss, B. Uniformly recurrent subgroups. In *Recent Trends in Ergodic Theory and Dynamical Systems (Contemporary Mathematics, 631)*. AMS, Providence; American Mathematical Society: Providence, RI, USA, 2015; pp. 63–75.
28. Matte Bon, N.; Tsankov, T. Realizing uniformly recurrent subgroups. *arXiv* **2017**, arXiv:1702.07101.
29. Protasov, I.V. Topological dualizms of locally compact abelian groups. *Ukr. Math. J.* **1977**, *29*, 625–631.
30. Protasov, I.V. Topological groups with compact lattice of closed subgroups. *Sib. Math. J.* **1979**, *20*, 378–385. [CrossRef]
31. Protasov, I.V. 0-dimensional groups with compact space of subgroups. *Math. Zamet.* **1985**, *37*, 483–490. [CrossRef]
32. Protasov, I.V. Compacts in the space of subgroups of a topological groups. *Ukr. Math. J.* **1986**, *38*, 600–605. [CrossRef]
33. Protasov, I.V. Topological groups with σ-compact spaces of subgroups. *Ukr. Math. J.* **1985**, *37*, 93–98. [CrossRef]
34. Protasov, I.V.; Saryev, A. Topological abelian groups with locally compact lattice of closed subgroups. *Dopov. AN Ukrain. SSR. Ser. A* **1980**, *N3*, 29–32.
35. Protasov, I.V. Limits of compact subgroups of a topological groups. *Dopov. AN Ukrain. SSR. Ser. A* **1986**, *N5*, 64–66.
36. Panasyuk, S.P. Metrizability in the space of subgroups of a Lie group. *Ukr. Math. J.* **1990**, *42*, 351–355. [CrossRef]
37. Panasyuk, S.P. Normality in the space of subgroups of a Lie group. *Ukr. Math. J.* **1990**, *42*, 786–788. [CrossRef]
38. Piskunov, A.G. Reconstruction of the Vectoris topology by compacts in the space of closed subgroups. *Ukr. Math. J.* **1990**, *42*, 789–794. [CrossRef]
39. Bourbaki, N. *Éléments de Matématique. Fascicule II. Livre III: Topologie Générale. Chapitre 1: Structures Topologiques. Chapitre 2: Structures Uniformes*; Hermann: Paris, France, 1940.
40. Protasov, I.V.; Saryev A. Bourbaki spaces of topological groups. *Ukr. Math. J.* **1990**, *42*, 542–549 [CrossRef]
41. Romaguera, S.; Sanchis, M. Completeness of hyperspaces of topological groups. *J. Pure Appl. Algebr.* **2000**, *149*, 287–293. [CrossRef]
42. Protasov, I.V.; Saryev A. Semigroup of closed subsets of a topological group, Izv. Akad. Nauk TadzhSSR. Ser. Fiz.-Tekh. Nauk **1988**, *3*, 21–25.
43. Itzkowitz, G.L. Continuous measures, Baire category, and uniformly continuous functions in topological groups. *Pac. J. Math.* **1974**, *54*, 115–125. [CrossRef]
44. Protasov, I.V. Functionally balanced groups. *Math. Acad. Sci. USSR* **1991**, *49*, 87–90. [CrossRef]
45. Protasov, I.V. Order convergence in the lattice of subgroups of a topological group. *Izv. Vysš. Učebn. Zav. Matematika* **1980**, *9*, 25–29.
46. Scheiderer, C. Topologies on subgroup lattice of a compact group. *Topol. Appl.* **1986**, *23*, 183–191. [CrossRef]
47. Protasov, I.V.; Stukotilov, V.S. Ochan topologies on a space of closed subgroups. *Ukr. Math. J.* **1989**, *41*, 1337–1342. [CrossRef]
48. Protasov, I.V. On topologies on lattices of subgroups. *Dopov. AN Ukr. SSR Ser. A* **1981**, *N11*, 29–32.

49. Roe, J. *Lectures on Coarse Geometry*; AMS University Lecture Ser. No. 31; American Mathematical Society: Providence, RI, USA, 2003.
50. Protasov, I.V.; Zarichnyi, M. *General Asymptopogy*; Mathematical Studies Monograph Series 12; VNTL: Lviv, Ukraine, 2007.
51. Dikranjan, D.; Protasov, I.; Zava, N. Hyperballeans of groups. *Topol. Appl.* **2018**, in press.

axioms

MDPI

Article

No Uncountable Polish Group Can be a Right-Angled Artin Group

Gianluca Paolini [1,*] and Saharon Shelah [1,2]

[1]   Einstein Institute of Mathematics, The Hebrew University of Jerusalem, Edmond J. Safra Campus, Givat Ram, Jerusalem 91904, Israel; shelah@math.huji.ac.il

[2]   Department of Mathematics, The State University of New Jersey, Hill Center-Busch Campus, Rutgers, 110 Frelinghuysen Road, Piscataway, NJ 08854-8019, USA

*   Correspondence: gianluca.paolini@mail.huji.ac.il; Tel.: +972-2-658-4103

Academic Editor: Sidney A. Morris
Received: 28 March 2017; Accepted: 4 May 2017; Published: 11 May 2017

Abstract: We prove that if G is a Polish group and A a group admitting a system of generators whose associated length function satisfies: (i) if $0 < k < \omega$, then $lg(x) \leq lg(x^k)$; (ii) if $lg(y) < k < \omega$ and $x^k = y$, then $x = e$, then there exists a subgroup G^* of G of size $\mathfrak{b}$ (the bounding number) such that G^* is not embeddable in A. In particular, we prove that the automorphism group of a countable structure cannot be an uncountable right-angled Artin group. This generalizes analogous results for free and free abelian uncountable groups.

Keywords: descriptive set theory; polish group topologies; right-angled Artin groups

---

In a meeting in Durham in 1997, Evans asked if an uncountable free group can be realized as the group of automorphisms of a countable structure. This was settled in the negative by Shelah [1]. Independently, in the context of descriptive set theory, Becher and Kechris [2] asked if an uncountable Polish group can be free. This was also answered negatively by Shelah [3], generalizing the techniques of [1]. Inspired by the question of Becher and Kechris, Solecki [4] proved that no uncountable Polish group can be free abelian. In this paper, we give a general framework for these results, proving that no uncountable Polish group can be a right-angled Artin group (see Definition 1). We actually prove more:

Theorem 1. *Let $G = (G, d)$ be an uncountable Polish group and A a group admitting a system of generators whose associated length function satisfies the following conditions:*

(i)   if $0 < k < \omega$, then $lg(x) \leq lg(x^k)$;
(ii) if $lg(y) < k < \omega$ and $x^k = y$, then $x = e$.

Then G is not isomorphic to A; in fact, there exists a subgroup G^* of G of size $\mathfrak{b}$ (the bounding number) such that G^* is not embeddable in A.

After the authors proved Theorem 1, they discovered that the impossibility to endow groups A as in Theorem 1 with a Polish group topology follows from an old important result of Dudley [5]. In fact, Dudley's work implies more strongly that we cannot even find a homomorphism from a Polish group G into A. Apart from the fact that the claim about G^* in Theorem 1 is of independent interest and not subsumed by Dudley's work, our focus here is on techniques; i.e., the crucial use of the Compactness Lemma of [3]. This powerful result has a broad scope of applications, and is used by the authors in a work in preparation [6] to deal with classes of groups not covered by Theorem 1 or Dudley's work, most notably the class of right-angled Coxeter groups (see Definition 1).

Proof of Theorem 1. Let $\zeta = (\zeta_n)_{n<\omega} \in \mathbb{R}^\omega$ be such that $\zeta_n < 2^{-n}$, for every $n < \omega$, and $\bar{g} = (g_n)_{n<\omega} \in G^\omega$ such that $g_n \neq e$ and $d(g_n, e) < \zeta_n$, for every $n < \omega$. Let Λ be a set of power

$\mathfrak{b}$ of increasing functions $\eta \in \omega^\omega$ which is unbounded with respect to the partial order of eventual domination. For transparency, we also assume that for every $\eta \in \Lambda$ we have $\eta(0) > 0$. For $\eta \in \Lambda$, define the following set of equations:

$$\Gamma_\eta = \{x_{n+1}^{\eta(n)} = x_n g_n : n < \omega\}.$$

By (3.1, [3]), for every $\eta \in \Lambda$, Γ_η is solvable in G. Let $\bar{b}_\eta = (b_{\eta,n})_{n<\omega}$ witness it; i.e.,

$$\bar{b}_\eta \in G^\omega \quad \text{and} \quad \bigwedge_{n<\omega} b_{\eta,n+1}^{\eta(n)} = b_{\eta,n} g_n.$$

Let G^* be the subgroup of G generated by $\{g_n : n < \omega\} \cup \{b_{\eta,n} : \eta \in \Lambda, n < \omega\}$. Towards contradiction, suppose that π is an embedding of G^* into A, and let S be a system of generators for A whose associated length function $lg_S = lg$ satisfies conditions (i) and (ii) of the statement of the theorem. For $\eta \in \Lambda$ and $n < \omega$, let:

$$\pi(g_n) = g_n', \quad \pi(b_{\eta,n}) = c_{\eta,n} \quad \text{and} \quad m_*(\eta) = lg(c_{\eta,0}).$$

Now, m_* is a function from Λ to ω and so there exists unbounded $\Lambda_1 \subseteq \Lambda$ such that for every $\eta \in \Lambda_1$ the value $m_*(\eta)$ is a constant m_*. Fix such a Λ_1 and m_*, and let $f_1, f_2 \in \omega^\omega$ increasing satisfying the following:

(1) $f_1(n) > lg(g_n')$;
(2) $f_2(n) = (m_* + 1) + \sum_{\ell<n} f_1(\ell)$.

Claim 1. *For every $\eta \in \Lambda_1$, $lg(c_{\eta,n}) < f_2(n)$.*

Proof. By induction on $n < \omega$. The case $n = 0$ is clear by the choice of f_1 and f_2. Let $n = m + 1$. Because of assumption (i) on A, the choice of Λ_1, and the choice of f_1 and f_2, we have:

$$\begin{aligned} lg(c_{\eta,n}) &\leq lg(c_{\eta,n}^{\eta(m)}) \\ &= lg(c_{\eta,m} g_m') \\ &\leq lg(c_{\eta,m}) + lg(g_m') \\ &< f_2(m) + f_1(m) \\ &= f_2(n). \end{aligned}$$

$\square$

Now, by the choice of Λ_1, we can find $\eta \in \Lambda_1$ and $n < \omega$ such that $\eta(n) > f_2(n+2)$. Notice then that by the claim above and the choice of f_1 and f_2, we have:

$$\eta(n) > f_2(n+1) = f_2(n) + f_1(n) > lg(c_{\eta,n}) + lg(g_n') \geq lg(c_{\eta,n} g_n'), \tag{1}$$

$$\eta(n) > f_2(n+2) \geq f_1(n+1) > lg(g_{n+1}'). \tag{2}$$

Thus, by (1) and the fact that $c_{\eta,n+1}^{\eta(n)} = c_{\eta,n} g_n'$, using assumption (ii), we infer that $c_{\eta,n+1} = e$. Hence,

$$c_{\eta,n+2}^{\eta(n+1)} = c_{\eta,n+1} g_{n+1}' = g_{n+1}'.$$

Furthermore, if $\eta(n+1) > lg(g_{n+1}')$, then again by assumption (ii), we have that $c_{\eta,n+2} = e$, and so $c_{\eta,n+2}^{\eta(n+1)} = g_{n+1}' = e$, which contradicts the choice of $(g_n)_{n<\omega}$. Hence, $\eta(n) < \eta(n+1) \leq lg(g_{n+1}')$, contradicting (2). It follows that the embedding π from G^* into A cannot exist. $\square$

Definition 1. *Given a graph* $\Gamma = (E, V)$, *the* right-angled Artin group $A(\Gamma)$ *is the group with presentation:*

$$\Omega(\Gamma) = \langle V \mid ab = ba : aEb \rangle.$$

If in the presentation $\Omega(\Gamma)$, *we ask in addition that all the generators are involutions, then we speak of* right-angled Coxeter groups $C(\Gamma)$.

Thus, for Γ, a graph with no edges (resp. a complete graph) $A(\Gamma)$ is a free group (resp. a free abelian group).

Definition 2. *Let* $A(\Gamma)$ *be a right-angled Artin group and* lg *its associated length function. We say that an element* $g \in A(\Gamma)$ *is cyclically reduced if it cannot be written as* $g = hfh^{-1}$ *with* $lg(g) = lg(f) + 2$.

Fact 1. *Let* $A(\Gamma)$ *be a right-angled Artin group,* lg *its associated length function, and* $g \in A(\Gamma)$. *Then:*

(1) g *can be written as* hfh^{-1} *with* f *cyclically reduced and* $lg(g) = lg(f) + 2lg(h)$;
(2) *if* $0 < k < \omega$ *and* f *is cyclically reduced, then* $lg(f^k) = klg(f)$;
(3) *if* $0 < k < \omega$ *and* $g = hfh^{-1}$ *is as in (1), then* $lg(hfh^{-1})^k = klg(f) + 2lg(h)$.

Proof. Item (1) is proved in (Proposition on p. 38, [7]). The rest is folklore. $\quad\square$

Corollary 1. *No uncountable Polish group can be a right-angled Artin group.*

Proof. By Theorem 1 it suffices to show that for every right-angled Artin group $A(\Gamma)$ the associated length function lg satisfies conditions (i) and (ii) of the theorem, but by Fact 1, this is clear. $\quad\square$

As is well known, the automorphism group of a countable structure is naturally endowed with a Polish topology which respects the group structure, hence:

Corollary 2. *The automorphism group of a countable structure cannot be an uncountable right-angled Artin group.*

As already mentioned, the situation is different for right-angled Coxeter groups; in fact, the structure M with ω many disjoint unary predicates of size 2 is such that $Aut(M) = (\mathbb{Z}_2)^\omega$; i.e., $Aut(M)$ is the right-angled Coxeter group on K_c (a complete graph on continuum many vertices). Notice that in this group for any $a \neq b \in K_c$, we have:

(i) $(ab)^2 = 1$;
(ii) $lg(ab) = 2 < 3$, $(ab)^3 = ab$ and $ab \neq e$.

Acknowledgments: Partially supported by European Research Council grant 338821. No. 1112 on Shelah's publication list.

Author Contributions: Gianluca Paolini and Saharon Shelah contribute equally to this manuscript.

Conflicts of Interest: The authors declare no conflict of interest.

References

1. Shelah, S. A Countable Structure Does Not Have a Free Uncountable Automorphism Group. *Bull. Lond. Math. Soc.* **2003**, 35, 1–7.
2. Becker, H.; Kechris, A.S. *The Descriptive Set Theory of Polish Group Actions*; London Math. Soc. Lecture Notes Ser. 232; Cambridge University Press: Cambridge, UK, 1996.
3. Shelah, S. Polish Algebras, Shy From Freedom. *Israel J. Math.* **2011**, 181, 477–507.
4. Solecki, S. Polish Group Topologies. In *Sets and Proofs*; London Math. Soc. Lecture Note Ser. 258; Cambridge University Press: Cambridge, UK, 1999.

5.   Dudley, R.M. *Continuity of Homomorphisms. Duke Math. J.* **1961**, *28*, 34–60.

6.   Paolini, G.; Shelah, S. Polish Topologies for Graph Products of Cyclic Groups. In Preparation.

7.   Servatius, H. Automorphisms of Graph Groups. *J. Algebra* **1989**, *126*, 34–60.

axioms

MDPI

Article

Computing the Scale of an Endomorphism of a totally Disconnected Locally Compact Group

George A. Willis

School of Mathematical and Physical Sciences, University of Newcastle, University Drive,
Callaghan NSW 2308, Australia; george.willis@newcastle.edu.au; Tel.: +61-2-4921-5666; Fax: +61-2-4921-6898

Received: 28 August 2017; Accepted: 9 October 2017; Published: 20 October 2017

Abstract: The scale of an endomorphism of a totally disconnected, locally compact group G is defined and an example is presented which shows that the scale function is not always continuous with respect to the Braconnier topology on the automorphism group of G. Methods for computing the scale, which is a positive integer, are surveyed and illustrated by applying them in diverse cases, including when G is compact; an automorphism group of a tree; Neretin's group of almost automorphisms of a tree; and a p-adic Lie group. The information required to compute the scale is reviewed from the perspective of the, as yet incomplete, general theory of totally disconnected, locally compact groups.

Keywords: locally compact group; endomorphism; scale; tree; Neretin's group; Thompson's group; p-adic Lie group

1. Introduction

Let G be a totally disconnected, locally compact (t.d.l.c.) group. A fundamental theorem about t.d.l.c. groups, proved by van Dantzig in the 1930s, see [1] and ([2] Theorem II.7.7), asserts that G has a base of neighbourhoods of the identity consisting of compact open subgroups. These subgroups are important for the definition of the *scale* of endomorphisms $\alpha : G \to G$, which is a positive integer gauging the action of α. The precise definition is given in Section 2. Computing the scale is a problem which, as will be seen in examples, depends very much on a description of the group G and of the endomorphism α (the examples in fact treat inner automorphisms), and one way to evaluate our understanding of general t.d.l.c. groups is whether we can carry out this computation.

Sections 3 and 4 describe two approaches to computing the scale and use them in examples. The first approach is the 'spectral radius formula' given by Rognvaldur Möller in [3], and the second uses the structure theorem given in [4,5] for compact open subgroups at which the scale is attained. One of the examples in Section 4 is a p-adic Lie group and it is seen that the scale may be computed in terms of eigenvalues in the Lie algebra and minimising subgroups in terms of eigenvectors. This observation, together with the spectral radius formula, results about the scale such as in [6], and applications of scale techniques to answer questions about t.d.l.c. groups which are answered in the connected case with approximation by Lie group methods [7,8], indicate that scale techniques may substitute for Lie methods in some circumstances. Motivation for developing improved methods for computing the scale is provided by these considerations.

Throughout the paper, End(G) will denote the monoid of continuous endomorphisms of G; Aut(G) will be the group of automorphisms of G; and Inn(G) the group of inner automorphisms of G.

2. The Scale of an Endomorphism

The scale was defined first for inner automorphisms of a t.d.l.c. group in [4,5] and the definition was extended to endomorphisms in [9]. For this definition, note that, if V is a compact open subgroup of G and α a continuous endomorphism, then $\alpha(V) \cap V$ is an open subgroup of the compact group $\alpha(V)$ and hence the index $[\alpha(V) : V \cap \alpha(V)]$ is finite.

Definition 1. *Let* $\alpha \in End(G)$. *The* scale *of* α *is the positive integer*

$$s(\alpha) := \min \{[\alpha(V) : V \cap \alpha(V)] \mid V \leq G \text{ compact and open}\}.$$

Any V at which the minimum is attained is minimising *for* α.

The scale and related concepts have been used in papers such as [7,8,10] to answer questions concerning t.d.l.c. groups. Many applications apply results to the *scale function on G* induced by the conjugation map $G \to Inn(G)$. For this scale function, we have the following, which is proved in [4].

Theorem 1. *The scale function* $s : G \to \mathbb{N}$ *is continuous for the group topology on G and the discrete topology on* $\mathbb{N}$.

This theorem is used in [10] to answer a question of K. H. Hofmann about the structure of t.d.l.c. groups. Ideas from the proof of the theorem have also been used to answer questions about the Chabauty space of closed subgroups of a t.d.l.c. group in [11,12].

It is a natural question whether the scale function is continuous with respect to the group topology with which $Aut(G)$ is usually equipped, namely, the Braconnier topology, see [13]. That is not the case, however, as may be seen as follows.

Example 1. *Let* $G = (\mathbb{K}((t))^2, +)$ *be the additive group of the 2-dimensional vector space over the field* $\mathbb{K}((t))$ *of formal Laurent series over the finite field* $\mathbb{K}$ *and let* α *be the automorphism* $\alpha(f, g) = (tf, t^{-1}g)$. *Then,* $s(\alpha) = |\mathbb{K}|$. *Define, for each* $n \in \mathbb{Z}$, $\alpha_n(f, g) = (f^{(n)}, g^{(n)})$ *where*

$$f^{(n)} = \sum_{k=-\infty}^{n} f_{k-1} t^k + \sum_{k=n+1}^{\infty} f_k t^k \quad and \quad g^{(n)} = \sum_{k=-\infty}^{n-1} g_{k+1} t^k + f_n t^n + \sum_{k=n+1}^{\infty} g_k t^k.$$

Then, it may be verified that $\alpha_n \to \alpha$ *in the Braconnier topology as* $n \to \infty$ *but* $s(\alpha_n) = 1$ *for every* n.

The lack of continuity of the scale might be remedied by considering the coarsest group topology, $\mathcal{T}$ say, on $Aut(G)$ finer than the Braconnier topology and for which the scale is continuous. (It has been pointed out by Christian Rosendal that, when G is second countable, $\mathcal{T}$ cannot be Polish if it is strictly finer than the Braconnier topology.) It would be desirable to have a more direct definition of $\mathcal{T}$. Even more desirable would be to have a directly defined topology on $End(G)$ for which $\mathcal{T}$ is the subspace topology on $Aut(G)$. In analogy with the ring of operators on a normed space and its open group of invertible operators, we might also ask whether $Aut(G)$ open in this topology?

3. Möller's "Spectral Radius" Formula

The scale may be calculated without finding a minimising subgroup.

Theorem 2. *[Spectral radius] Let V be any compact open subgroup of G and* α *be in* $End(G)$. *Then, the scale of* α *is equal to*

$$s(\alpha) = \lim_{n \to \infty} [\alpha^n(V) : \alpha^n(V) \cap V]^{\frac{1}{n}}.$$

This formula is referred to as the "spectral radius formula" because of the similarity with the formula of the same name for linear operators on normed spaces. It turns out that the scale of an element x in a t.d.l.c. group G is exactly the spectral radius of the operator of translation by x on a certain normed convolution algebra on G, see [14]. The formula was proved for automorphisms by R. G. Möller in ([3] Theorem 7.7) and extended to endomorphisms in [9, Proposition].

The spectral radius formula is illustrated by the next two examples.

Example 2. *Let* $\mathcal{T}_{q+1}$ *be the regular tree in which every vertex has valency* $q + 1$ *and let* $G = \text{Aut}(\mathcal{T}_{q+1})$. *It may be seen that* G *is a topological group under the topology of pointwise convergence on vertices and that for each vertex,* v, *in* $\mathcal{T}_{q+1}$ *the stabiliser* $\text{stab}_G(v)$ *is open. Since, moreover,* $\text{stab}_G(v)$ *isomorphic as a topological group to the iterated wreath product* $S_{q+1} \wr S_q \wr S_q \wr \dots$ *and is therefore profinite, it follows that* G *is a locally compact group. It is a totally disconnected group because, for each pair* $x \neq y \in G$ *we may choose* $v \in V(\mathcal{T}_{q+1})$ *with* $x.v \neq y.v$ *and then the sets* $\mathcal{U}_1 = \{g \in G \mid g.v = x.v\}$ *and* $\mathcal{U}_2 = \{g \in G \mid g.v \neq x.v\}$ *are an open partition of* G *with* $x \in \mathcal{U}_1$ *and* $y \in \mathcal{U}_2$.

Every automorphism of G *is inner. This may be shown, see [15], by observing that* $\{\text{stab}_G(v) \mid v \in V(\mathcal{T}_{q+1})\}$ *is a set of maximal compact open subgroups of* G *on which each automorphism,* α *say, acts, and that the tree* $\mathcal{T}_{q+1}$ *may be reconstructed from this set of subgroups. The action of* α *on* $\{\text{stab}_G(v) \mid v \in V(\mathcal{T}_{q+1})\}$ *thus induces an automorphism,* x_α, *of* $\mathcal{T}_{q+1}$ *and it may be seen that* α *is equal to the inner automorphism of conjugation by* x_α.

Consider $x \in G$ *and the inner automorphism* $\alpha_x : g \mapsto x g x^{-1}$. *Let* $U = G_v$ *for some vertex* v. *As described in [15], there are two cases:* x *could have finite orbits in* $\mathcal{T}_{q+1}$, *in which case it is called* elliptic; *or* x *could have infinite orbits in* $\mathcal{T}_{q+1}$, *in which case it is called* hyperbolic.

x is elliptic. In this case, $\{\alpha_x^n(U) \mid n \geq 0\} = \{\text{stab}_G(x^n.v) \mid n \geq 0\}$ *is finite and hence the set of indices* $[\alpha_x^n(U) : \alpha_x^n(U) \cap U]$ *is bounded. Then,*

$$s(x) = \lim_{n \to \infty} [\alpha_x^n(U) : \alpha_x^n(U) \cap U]^{\frac{1}{n}} = 1.$$

x is hyperbolic. In this case, x *is a translation along a geodesic path* ℓ, *where a vertex* w *is on* ℓ *if the distance from* w *to* $x.w$ *is a minimum. Let this minimum distance be* d, *so that* x *translates* ℓ *by distance* d. *Let* w *be the closest vertex on* ℓ *to the given vertex* v. *Then,* $x.w$ *is the closest vertex on* ℓ *to* $x.v$ *and the path from* $v \to w \to x.w \to x.v$ *is the shortest path from* v *to* $x.v$ *and has length* $d + 2c$, *where* c *is the distance from* v *to* ℓ. *Since the index* $[\text{stab}_G(u_1) : \text{stab}_G(u_1) \cap \text{stab}_G(u_2)]$ *is equal to* $(q+1)q^{d(u_1,u_2)-1}$ *for any two vertices* u_1 *and* u_2, *as may be seen by an application of the Orbit-Stabiliser Theorem, it follows that*

$$[\alpha_x^n(U) : \alpha_x^n(U) \cap U] = [\text{stab}_G(x^n.v), \text{stab}_G(x^n.v) \cap \text{stab}_G(v)] = (q+1)q^{dn+2c-1}.$$

Hence

$$s(x) = \lim_{n \to \infty} \left((q+1)q^{dn+2c-1} \right)^{\frac{1}{n}} = q^d.$$

Example 3. *Let* $G = F^{\mathbb{Z}}$, *where* F *is a non-trivial finite group, and equip* G *with the product topology and the pointwise product. Then,* G *is a compact group. Hence every endomorphism of* G *has scale 1 because* G *itself is a compact subgroup invariant under the endomorphism.*

Let α *the shift automorphism* $\alpha(f)_n = f_{n+1}$. *Consider the subgroup*

$$U_K = \left\{ f \in F^{\mathbb{Z}} \mid f(k) = 1_F \text{ for } 0 \leq k < K \right\}.$$

An easy calculation shows that

$$[\alpha^n(U_K) : \alpha^n(U_K) \cap U_K] = \begin{cases} |F|^n, & \text{if } 0 \leq n \leq K \\ |F|^K, & \text{if } n > K \end{cases}.$$

Hence we find, as expected, that

$$s(\alpha) = \lim_{n \to \infty} |F|^{\frac{K}{n}} = 1.$$

4. Identifying Minimising Subgroups

Minimising subgroups have a structural characterisation which may be used to calculate the scale. Moreover, the proof of this characterisation involves a procedure for finding minimising subgroups.

The characterisation and the procedure for finding minimising subgroups will now be described and then illustrated in several examples.

The characterisation of minimising subgroups for $\alpha \in \text{End}(G)$ involves the following two subgroups defined for any compact and open subgroup V of G.

$$V_+ = \{x \in V \mid \exists \{x_n\}_{n \in \mathbb{N}} \subset V \text{ with } x_0 = x \text{ and } x_n = \alpha(x_{n+1}) \text{ for each } n \in \mathbb{N}\} \qquad (1)$$
$$V_- = \{x \in V \mid \alpha^n(x) \in V \text{ for all } n \in \mathbb{N}\}.$$

The sequence $\{x_n\}_{n \in \mathbb{N}}$ appearing in the definition of V_+ is a "history" of x as α is iterated and the condition for x to be in V_+ is that this history is contained in V. This history need not be unique because α need not be injective, and it is not required that all histories of x lie in V. The condition for x to be in V_- is that the "future" of x when α is iterated should lie in V.

In what follows, it is important to note that $\alpha(V_+) \geq V_+$. Hence, $\bigcup_{n \in \mathbb{N}} \alpha^n(V_+)$ is an increasing union of subgroups of G and is therefore itself a subgroup.

Theorem 3 (The structure of minimising subgroups). *Let $\alpha \in \text{End}(G)$ and V be a compact open subgroup of G. Then, V is minimising for α if and only if*

TA: $V = V_+ V_-$
TB1: $V_{++} := \bigcup_{n \in \mathbb{N}} \alpha^n(V)$ *is closed and*
TB2: *the sequence of integers $\{[\alpha^{n+1}(V_+) : \alpha^n(V_+)] \mid n \in \mathbb{N}\}$ is constant.*

In this case, $s(\alpha) = [\alpha(V_+) : V_+]$.

Definition 2. *A subgroup V is* tidy above *for α, if it satisfies condition TA;* tidy below *if it satisfies conditions TB1 and TB2; and is* tidy *if it is tidy above and below.*

The condition TB2 is redundant if α is injective because it follows from the other conditions in that case. Hence, this condition did not appear in [4], which deals with automorphisms only.

4.1. The Tidying Procedure

This section describes a three-step procedure which takes as input a general compact open subgroup, U, and produces a subgroup tidy for α. The procedure is given effect by the following three propositions.

Proposition 1 (Step 1). *Let $U \leq G$ be compact and open and $\alpha \in \text{End}(G)$. Then, there is $N \in \mathbb{N}$ such that the subgroup $U_{-N} = \bigcap_{n=0}^{N} \alpha^{-n}(U)$ is tidy above for α. For this N, we have*

$$[\alpha(U_{-N}) : \alpha(U_{-N}) \cap U_{-N}] \leq [\alpha(U) : \alpha(U) \cap U].$$

Remark 1. *The proof of the proposition involves forming the decreasing sequence of subgroups $\{U_k\}_{k \in \mathbb{N}}$, where*

$$U_k = \{u \in U \mid \exists u_0, u_1, \ldots, u_k \in U \text{ with } u_n = \alpha(u_{n+1}) \text{ for } 0 \leq n < k \text{ and } u_0 = u\}. \qquad (2)$$

Then, $\bigcap_{k \in \mathbb{N}} U_k = U_+$ and N is the first k such that $U_k \subset U_+(U \cap \alpha^{-1}(U))$. That such k exists follows from compactness of U_k and the fact that $U_+(U \cap \alpha^{-1}(U))$ is an open neighbourhood of U_+. This step involves cutting down U to the subgroup U_{-N} and motivates the name 'tidy above' for the factorisation property satisfied by U_{-N}.

The next two steps involve ensuring that the group is 'tidy below' by including in it the compact subgroup identified in the next proposition.

Proposition 2 (Step 2). *Suppose that $V \leq G$ is tidy above for α and define*

$$\mathcal{L}_V = \{x \in G \mid \exists m, n \in \mathbb{N} \text{ with } x \in \alpha^m(V_+) \text{ and } \alpha^n(x) \in V_-\}.$$

Then, $\overline{\mathcal{L}_V}$ is compact and α-stable.

Local compactness of G is again important in the proof of this proposition.

The third proposition combines V with $\mathcal{L}_V$ to form a subgroup that is tidy for α. It is not enough to simply multiply the two subgroups because that might not be a subgroup. The subgroup generated by V and $\mathcal{L}_V$ might not be compact, and so that method of combining the subgroups will not work either.

Proposition 3 (Step 3). *Suppose that $V \leq G$ is tidy above for α and define*

$$\tilde{V} = \{x \in V \mid x\mathcal{L}_V \subset \mathcal{L}_V V\} \text{ and } W = \tilde{V}\mathcal{L}_V.$$

Then W is "a compact open subgroup of G that is tidy for α" and

$$[\alpha(W) : \alpha(W) \cap W] \leq [\alpha(V) : \alpha(V) \cap V].$$

4.2. Tidy Subgroups Are Minimising and Conversely

These three steps take a general compact open subgroup and modify it to produce a tidy subgroup. The next result implies that this subgroup is minimising for α.

Theorem 4. *The index $[\alpha(W) : \alpha(W) \cap W]$ is the same for all compact open subgroups tidy for α.*

To prove the claim that tidy subgroups are minimising, suppose that the compact open subgroup U is minimising and apply the tidying procedure to U. Then, the subgroup W so produced is tidy and $[\alpha(W) : \alpha(W) \cap W] \leq [\alpha(U) : \alpha(U) \cap U]$. Since U was already minimising, we conclude that W is minimising and hence so are all subgroups tidy for α.

That all minimising subgroups are tidy may be seen by noting that the inequalities in Propositions 1 and 3 are equalities if and only if the group already satisfies the relevant tidiness condition. Hence, if U is already minimising, the tidying procedure does not alter it and U is therefore tidy.

4.3. Tidy Subgroups and the Scale in Examples

Theorem 3 and the notion of tidy subgroup will now be illustrated by using them to compute the scale for the same automorphisms as in the previous section, as well as for some additional examples.

Example 2 (Revisited), let $G = \text{Aut}(T_{q+1})$ and $\alpha = \alpha_x$ be an automorphism as before. The tidying procedure will be applied with $U = \text{stab}_G(v)$ for an arbitrary vertex v in T_{q+1}. For this, note that $\alpha_x^n(U) = \text{stab}_G(x^n.v)$.

Since α_x is an automorphism, and is in particular injective, the subgroup U_k defined as in Equation (2), is equal to $\bigcap_{j=0}^k \alpha^j(U)$ and the subgroup U_+ defined as in Equation (1) is equal to $\bigcap_{j=0}^\infty \alpha^j(U)$. Hence, U_k is the fixator of the vertices $v, x.v, \ldots, x^k.v$ and U_+ is the fixator of the vertices $x^n.v$, $n \in \mathbb{N}$. As before, the cases when x is elliptic and when it is hyperbolic are treated separately.

x is elliptic. The orbit $x^n.v$, $n \in \mathbb{Z}$ is finite and so there is k such that $\{v, x.v, \ldots, x^k.v\}$ is equal to this orbit. Choosing $N = k$, we then have that $U_+ = U_k$ and it is easily seen that U_- is equal to this subgroup as well. Hence, putting

$$V = U_k = \text{stab}_G(v) \cap \text{stab}_G(x.v) \cap \cdots \cap \text{stab}_G(x^k.v),$$

we have that $\alpha_x(V) = V = V_+ = V_-$. Hence, V is tidy above. It follows that $V_{++} = V$ as well and is closed. Therefore, V is tidy for α_x. Of course,

$$[\alpha_x(V) : \alpha_x(V) \cap V] = [V : V] = 1.$$

x is hyperbolic. As before, let ℓ be the axis for x, suppose that x translates along ℓ through distance d, let c be the distance from v to ℓ, and let w be the vertex on ℓ that is closest to v. In addition, denote the neighbour of w closest to $x.v$ by w^+ and the neighbour of w closest to $x^{-1}.v$ by w^-.

The subgroup U itself is not tidy for α_x. To see this, note that since U_+ fixes all vertices $x^n.v$ with $n \geq 0$, it fixes w and w^+ as well. Similarly, U_- fixes w and w^-. Hence $U \neq U_+U_-$ if $v \neq w$ and U is not tidy above. The same conclusion holds even when $v = w$ because, while U acts as the full permutation group S_{q+1} on the $q+1$ neighbours of w, U_+ and U_- each fix one of the neighbours and S_{q+1} is not equal to the product of two such subgroups. We see too that, since x fixes every vertex on the path from v to $x.v$, the same calculation as in the earlier discussion yields that

$$[\alpha_x(U) : \alpha_x(U) \cap U] = (q+1)q^{d+2c-1}, \tag{3}$$

which is strictly greater than q^d. Hence, U is not minimising.

Step 1 The subgroup $U_{-1} = U \cap \alpha_x^{-1}(U)$ is tidy above however. Setting $V = U_{-1}$, we have that V fixes all vertices on the path from w to $x^{-1}.w$. Hence, the Tits independence property for $\text{Aut}(\mathcal{T}_{q+1})$ implies that $V = H_+H_-$, where H_+ and H_- are the fixators in V of the components $\mathcal{S}_+$ and $\mathcal{S}_-$ of $\mathcal{T}_{q+1}$ formed when the path from w to $x^{-1}.w$ is deleted and containing $x^{-1}.w$ and w respectively. Since $H_\pm \leq V_\pm$, it follows that V is tidy above. The index we are interested in may be calculated to be

$$[\alpha_x(V) : \alpha_x(V) \cap V] = \begin{cases} (q-1)q^{d+c}, & \text{if } v \notin \ell \\ q^d, & \text{if } v \in \ell \end{cases} \tag{4}$$

which is strictly less than that found in (3). However, it is not the minimum value calculated using the spectral radius formula unless v happens to lie on ℓ, or $q = 2$ and $c = 1$. The second and third steps of the tidying procedure must therefore be implemented to find a minimising subgroup.

Carrying out these steps will require labelling some more vertices of $\mathcal{T}_{q+1}$. Denote the set of $q+1$ neighbours of w by $N(w)$ and similarly for $x^{-1}.w$. Recall that w^+ and w^- are neighbours of w on ℓ, and denote by w^0 the neighbour of w lying on the path from v to w. Then $x^{-1}.w^+$, $x^{-1}.w^-$ and $x^{-1}.w^0$ are neighbours of $x^{-1}.w$: $x^{-1}.w^+$ is the neighbour closest to w; $x^{-1}.w^-$ is closest to $x^{-2}.w$; and $x^{-1}.w^0$ lies on the path from $x^{-1}.v$ to $x^{-1}.w$. The subgroup V found in the previous paragraph fixes all vertices on ℓ between w and $x^{-1}.w$ as well as the vertices on the paths joining v to w and $x^{-1}.v$ to $x^{-1}.w$, that is, all vertices on the path from v to $x^{-1}.v$. In particular, V acts on $N(w)$ by fixing w^0 and w^- and as the full symmetric group on $N(w) \setminus \{w^0, w^-\}$; and acts on $N(x^{-1}.w)$ by fixing $x^{-1}.w^0$ and $x^{-1}.w^+$ and as the full symmetric group on $N(x^{-1}.w) \setminus \{x^{-1}.w^0, x^{-1}.w^+\}$.

Step 2 The definition of $\mathcal{L}_V$ in Proposition 2 implies that y belongs to this subgroup if $\alpha_x^n(y) \in V$ for all but finitely many n. Hence, tree automorphisms in $\mathcal{L}_V$ fix all the vertices of ℓ and all vertices on paths joining $x^n.v$ to ℓ except for finitely many n. The action of $\mathcal{L}_V$ on $N(w)$ thus fixes w^+ and w^- and permutes vertices in $N(w) \setminus \{w^-, w^+\}$ arbitrarily; and its action on $N(x.w)$ fixes $x^{-1}.w^-$ and $x^{-1}.w^+$ and permutes vertices in $N(x.w) \setminus \{x^{-1}.w^+, x^{-1}.w^-\}$ arbitrarily. That the closure of $\mathcal{L}_V$ is compact as claimed in Proposition 2 may be seen by observing that this closure is the fixator of all vertices on ℓ.

Step 3 That the product $\mathcal{L}_V V$ is not a group may be seen by considering its action on $N(w)$. While V and $\mathcal{L}_V$ both fix w, we have that $\mathcal{L}_V$ fixes w^+ and w^- and acts as $\text{Sym}(N(w)) \setminus \{w^-, w^+\}$) on the remaining vertices in $N(w)$; and V fixes w^0 and w^- and acts as $\text{Sym}(N(w)) \setminus \{w^0, w^-\}$) on the remaining vertices; but the product

$$\mathrm{Sym}(N(w)) \setminus \{w^-, w^+\})\mathrm{Sym}(N(w) \setminus \{w^o, w^-\})$$

is not a subgroup of $\mathrm{Sym}(N(w))$. In the present example, $\langle \mathcal{L}_V, V \rangle$ is compact and equal to the fixator of the path from w to $x^{-1}.w$, which is tidy for α_x. We shall see, however, that the procedure described in Proposition 3 produces a different tidy subgroup. According to this procedure, define $\tilde{V} = \{g \in V \mid g\mathcal{L}_V \subset \mathcal{L}_V V\}$. To determine $\tilde{V}$, we apply the following lemma about finite permutation groups to the subgroups $\mathrm{Sym}(N(w)) \setminus \{w^o, w^-\})$ and $\mathrm{Sym}(N(w)) \setminus \{w^+, w^-\})$ of $\mathrm{Sym}(N(w)) \setminus \{w^-\})$.

Lemma 1. *Let S_q denote the permutation group $S_q = \mathrm{Sym}(\{1, 2, \ldots, q\})$. Then,*

$$\left\{ \pi \in stab_{S_q}(1) \mid \pi stab_{S_q}(q) \subset stab_{S_q}(q) stab_{S_q}(1) \right\} = stab_{S_q}(1, q).$$

Proof. Note that no permutation in $stab_{S_q}(q)stab_{S_q}(1)$ sends 1 to q. On the other hand, if $\pi \in stab_{S_q}(1)$ and does not fix q, then there is $j \in \{2, \ldots, q-1\}$ such that $\pi(j) = q$ and there is $\sigma \in stab_{S_q}(q)$ such that $\sigma(1) = j$. Then $\pi\sigma(1) = q$ and is not in $stab_{S_q}(q)stab_{S_q}(1)$. $\square$

It follows from Lemma 1 that

$$\tilde{V} \subset \{g \in V \mid g \text{ fixes } w^o, w^- \text{ and } w^+\} \cap \{g \in V \mid g \text{ fixes } x^{-1}.w^o, x^{-1}.w^- \text{ and } x^{-1}.w^+\}.$$

Since elements of V fix w^o, w^-, $x^{-1}.w^+$ and $x^{-1}.w^o$ already,

$$\tilde{V} \subset \{g \in V \mid g \text{ fixes } w^+ \text{ and } x^{-1}.w^-\}.$$

It is easily seen that all elements of V fixing w^+ and $x^{-1}.w^-$ belong to $\tilde{V}$. Hence $\tilde{V} = stab_V(w^+) \cap stab_V(x^{-1}.w^-)$. Then,

$$W = \mathcal{L}_V\tilde{V} = stab_G(w^+) \cap stab_G(x^{-1}.w^-),$$

that is, elements of W fix all vertices on the axis ℓ between $x^{-1}.w^-$ and w^+.

To compute the scale of α_x using this tidy subgroup W, observe that $\alpha_x(W)$ fixes all vertices on ℓ between w^- and $x.w^+$ and $\alpha_x(W) \cap W$ fixes all vertices between $x^{-1}.w^-$ and $x.w^+$. The distance on ℓ between $x^{-1}.w^-$ and w^- is d, and the orbit of $x^{-1}.w^-$ under $\alpha_x(W)$ therefore has order q^d. Hence,

$$s(\alpha_x) = [\alpha_x(W) : \alpha_x(W) \cap W] = q^d.$$

Regular trees are a particular type of *building*—see [16] for the definition—and automorphism groups of locally finite buildings are totally disconnected, locally compact groups. The calculation of the scale in terms of geometric data describing the building could also be carried out by a similar approach to that used for trees.

Example 3 (Revisited), let $G = F^{\mathbb{Z}}$ for some finite group F and α be the shift automorphism as before. It has already been remarked that G is compact and invariant under α and so $s(\alpha) = 1$. When the tidying procedure is applied with $U = G$ there is no change: in Step 1, we have $N = 0$ and so $V = U$ in the next step; in Step 2, $\mathcal{L}_V = V$; and in Step 3, $\tilde{V} = V$ and $W = V = U$.

The tidying procedure will be illustrated by applying it to the compact open subgroup

$$U_X = \left\{ f \in F^{\mathbb{Z}} \mid f(k) = 1_F \text{ for all } k \in X \right\},$$

where X is a finite subset of $\mathbb{Z}$. When $X = \{0, 1, \ldots, K-1\}$, we recover the subgroup U_K considered previously.

Step 1 Since α is an automorphism, the group $(U_X)_n$ defined in Proposition 1 is equal to $\bigcap_{k=0}^{n} \alpha^k(U_X)$ and the number N whose existence is guaranteed by the proposition depends on X.

In the case when $X = \{0, 1, \ldots, K - 1\}$ the subgroup $U_X = U_K$ is already tidy above and $N = 0$. To see this, note that

$$(U_K)_+ = \left\{ f \in F^{\mathbb{Z}} \mid f(k) = 1_F \text{ if } k < K \right\} \text{ and } (U_K)_- = \left\{ f \in F^{\mathbb{Z}} \mid f(k) = 1_F \text{ if } k \geq 0 \right\}.$$

In this case, we have $[\alpha(U_K) : \alpha(U_K) \cap U_K] = |F|$.

For another case, suppose that $X = \{0, 1, 5, 6, 7, 8\}$. Then,

$$\alpha^n(U_X) = \left\{ f \in F^{\mathbb{Z}} \mid f(k) = 1_F \text{ if } k \in X - n \right\}.$$

Hence, $\alpha(U_X) \cap U_X = \{ f \in F^{\mathbb{Z}} \mid f(k) = 1_F \text{ if } k \in X \cup (X - 1) \}$. In other words,

$$\alpha(U_X) = \left\{ f \in F^{\mathbb{Z}} \mid f(k) = 1_F \text{ if } k \in \{-1, 0, 4, 5, 6, 7\} \right\} \text{ and}$$
$$\alpha(U_X) \cap U_X = \left\{ f \in F^{\mathbb{Z}} \mid f(k) = 1_F \text{ if } k \in \{-1, 0, 1, 4, 5, 6, 7, 8\} \right\}$$

and $[\alpha(U_X) : \alpha(U_X) \cap U_X] = |F|^2$. Moreover,

$$(U_X)_+ = \left\{ f \in F^{\mathbb{Z}} \mid f(k) = 1_F \text{ if } k < 9 \right\} \text{ and } (U_K)_- = \left\{ f \in F^{\mathbb{Z}} \mid f(k) = 1_F \text{ if } k \geq 0 \right\}$$

and $(U_X)_+(U_X)_- \neq U_X$. Similar calculations apply for $(U_X)_1 = \alpha(U_X) \cap U_X$ and $(U_X)_2 = \alpha^2(U_X) \cap \alpha(U_X) \cap U_X$. However,

$$(U_X)_3 = \left\{ f \in F^{\mathbb{Z}} \mid f(k) = 1_F \text{ if } -3 \leq k \leq 8 \right\},$$

which is tidy above, and $[\alpha((U_X)_3) : \alpha((U_X)_3) \cap (U_X)_3] = |F|$.

Steps 2 and 3 For any subgroup U_X with $X \neq \varnothing$ we have

$$\mathcal{L}_{U_X} = \left\{ f \in F^{\mathbb{Z}} \mid f(k) = 1_F \text{ for all but finitely many } k \right\}$$

and $\mathcal{L}_{U_X} U_X = G$.

Lie groups over local fields are totally disconnected and locally compact as well, and the scale of elements in such groups, that is, of inner automorphisms of the groups, was computed by H. Glöckner in [17]. His were the first calculations of the scale for groups that went beyond the basic cases seen in the previous examples.

Example 4. *Let G be a Lie group over the field $\mathbb{Q}_p$ of p-adic numbers. Glöckner does not use the tidying procedure to find subgroups tidy for x (that is, for α_x) but instead describes V_+ and V_- directly in terms of the normal form of the Lie algebra automorphism Ad_x and calculates $s(x)$ in terms of eigenvalues of Ad_x (in a finite extension of $\mathbb{Q}_p$). This correspondence between the scale and tidy subgroups on one hand and eigenvalues and eigenspaces on the other is evidence that scale techniques are a substitute for Lie algebra techniques when studying t.d.l.c. groups that are not Lie groups over local fields.*

The main ideas in [17] may be sketched as follows. Assume that V is tidy for x, then V_{++} is closed and so is a Lie subgroup of G. Moreover, V_+ is an open subgroup of V_{++} and $s(x) = \Delta(I_x)$, where I_x is the automorphism of V_{++} induced by conjugation by x and Δ is the module function on automorphisms. The module of this automorphism of V_{++} is then equal to the module of the automorphism $L(I_x)$ induced on the Lie algebra of V_{++}. Glöckner then describes this Lie algebra as a subalgebra of $L(G)$ (the Lie algebra of G) in what he calls the contraction decomposition *of Ad_x. This decomposition applies to any linear automorphism ϕ of a finite-dimensional vector space, L, over a local field of characteristic 0, and expresses L as*

$$L = L_p \oplus L_0 \oplus L_m$$

where

$$L_p := \left\{ y \in L : \phi^{-k}(y) \to 0 \text{ as } k \to \infty \right\}$$

$$L_m := \left\{ y \in L : \phi^k(y) \to 0 \text{ as } k \to \infty \right\} \text{ and}$$

$$L_0 := \left\{ y \in L : \{ \| \phi^k(y) \| \}_{k \in \mathbb{Z}} \text{ is bounded} \right\}.$$

Glöckner bases this decomposition on a variation on ([18] Lemma 3.4). Applying it when L is the Lie algebra of G and $\phi = \mathrm{Ad}_x$, he finds that $L_p \oplus L_0$ is a subalgebra isomorphic to the Lie algebra of V_{++}. Using this decomposition, Glöckner shows in ([17] Corollary 3.6) that, if G is a Lie group over a local field, K, of characteristic 0 and x is in G, then

$$s(x) = \prod_{|\lambda_i| \geq 1} |\lambda_i|,$$

where λ_i are the roots of the characteristic polynomial of Ad_x in a splitting field, K', for this polynomial and $| \cdot |$ is the unique extension to K' of absolute value on K.

Glöckner thus reduces the computation of the scale to finding eigenvalues and avoids the need to find tidy subgroups. He gives more explicit formulæ for the scale on linear algebraic groups. The formulæ and their relation to the methods for computing the scale previously discussed may be illustrated with the case when the group is $\mathrm{GL}(2, \mathbb{Q}_p)$ and the element x has the property that its characteristic polynomial splits over $\mathbb{Q}_p$, in which case there is a basis for $\mathbb{Q}_p^2$ with respect to which x has a diagonal matrix with entries λ_1, λ_2, the eigenvalues of x. Consider the compact, open subgroup $U = \mathrm{GL}(2, \mathbb{Z}_p)$. Note that the condition that U is a group forces the determinant of each element of U to have p-adic absolute value equal to 1.

Applying powers of α_x to U yields that

$$\alpha_x^n(U) = \left\{ \begin{pmatrix} a & (\lambda_1 \lambda_2^{-1})^n b \\ (\lambda_1^{-1} \lambda_2)^n c & d \end{pmatrix} : a, b, c, d \in \mathbb{Z}_p, |ad - bc|_p = 1 \right\}$$

$$= \left\{ \begin{pmatrix} a & p^{-kn} b \\ p^{kn} c & d \end{pmatrix} : a, b, c, d \in \mathbb{Z}_p, |ad - bc|_p = 1 \right\}, \tag{5}$$

where $|\lambda_1 \lambda_2^{-1}| = p^k$. Thus, if $k = 0$, then $\alpha_x(U) = U$ and $s(\alpha_x) = s(x) = 1$. Suppose that $k > 0$. (The case when $k < 0$ is similar.) Then,

$$\alpha_x^n(U) \cap U = \left\{ \begin{pmatrix} a & b \\ p^{kn} c & d \end{pmatrix} : a, b, c, d \in \mathbb{Z}_p, |ad - bc|_p = |ad|_p = 1 \right\}$$

and it may be calculated that

$$[\alpha_x^n(U) : \alpha_x^n(U) \cap U] = (p+1)p^{kn-1}.$$

Hence, by Theorem 2, $s(x) = \lim_{n \to \infty} \left((p+1)p^{kn-1} \right)^{\frac{1}{n}} = p^k = |\lambda_1 \lambda_2^{-1}|$, which is the same value as given in the last example in ([17] Corollary 3.6).

The subgroup U is not tidy above for α_x when $k > 0$ because, as follows from (5),

$$U_+ = \left\{ \begin{pmatrix} a & b \\ 0 & d \end{pmatrix} : a, b, d \in \mathbb{Z}_p, |ad|_p = 1 \right\} \text{ and } U_- = \left\{ \begin{pmatrix} a & 0 \\ c & d \end{pmatrix} : a, c, d \in \mathbb{Z}_p, |ad|_p = 1 \right\}$$

and the element $u = \begin{pmatrix} 0 & 1 \\ 1 & 0 \end{pmatrix}$ belongs to U but not to $U_+ U_-$.

Step 1 *of the tidying procedure. The subgroup*

$$V = U \cap \alpha_x(U) = \left\{ \begin{pmatrix} a & b \\ p^k c & d \end{pmatrix} : a, b, c, d \in \mathbb{Z}_p, |ad|_p = 1 \right\}$$

may be verified to be tidy above by showing that every element of V is the product of an upper triangular and a lower triangular matrix. Proposition 1 thus holds with N = 1 in this case.

Steps 2 and 3 *It may also be verified that*

$$V_{++} = \left\{ \begin{pmatrix} a & b \\ 0 & d \end{pmatrix} : b \in \mathbb{Q}_p, a, d \in \mathbb{Z}_p, |ad|_p = 1 \right\},$$

which is a closed subgroup of $\mathrm{GL}(2, \mathbb{Q}_p)$*, and hence that V is also tidy below. That the tidying procedure terminates after the first step in this case is no accident: it is shown in ([19] Theorem 3.2) that that occurs for any automorphism α for which the contraction subgroup,*

$$\mathrm{con}(\alpha) = \{ y \in G : \alpha^n(y) \to 1_G \text{ as } n \to \infty \},$$

is closed and it is shown in ([18] Theorem 3.5) that that is always so for automorphisms of p-adic Lie groups.

    Glöckner has also calculated the scale in some cases for linear groups over a skew field, K, with positive characteristic, see [20]. He shows that, if x is a diagonalisable element in $\mathrm{GL}(n, K)$*,* $\mathrm{SL}(n, K)$*,* $\mathrm{PGL}(n, K)$ *or* $\mathrm{PSL}(n, K)$*, then the scale is given by the same formula as in the characteristic 0 case. He does so by writing down tidy subgroups for* α_x*. In particular, he shows that, if the diagonal entries in x are in order of decreasing modulus, then certain compact, open subgroups may be written as the product of their subgroup of upper triangular matrices with their subgroup of lower triangular matrices and that this implies tidiness above. Tidiness below is again satisfied automatically, as in the previous paragraph, because contraction subgroups for inner automorphisms are closed.*

    More is known about *p*-adic Lie groups than for general t.d.l.c. groups but a key question remains unanswered even for these groups. If an element *x* in a t.d.l.c. group satisfies that $s(x) = 1 = s(x^{-1})$, then subgroups tidy for *x* are normalised by *x* and, conversely, if *U* is normalised by *x*, then $s(x) = 1 = s(x^{-1})$. A t.d.l.c. group *G* is *uniscalar* if $s(x) = 1$ for every $x \in G$ and it is shown in [21], relying on a result in [22], that a *p*-adic Lie group that is compactly generated and uniscalar has a compact, open normal subgroup. There are uniscalar t.d.l.c. groups having no compact, open normal subgroups which are compactly generated, see [23], and which are topologically simple, see [24]. However, no examples are known of uniscalar t.d.l.c. groups which are topologically simple and compactly generated, or which are topologically simple (of necessity not compactly generated) and *p*-adic Lie.

    Another significant class of t.d.l.c. groups are the groups of *almost automorphisms* of trees introduced by Yu. Neretin, [25,26], and shown to be simple by C. Kapoudjian, [27]. Neretin groups are also studied in [28], where it is shown that they do not contain a lattice, and the notation used here conforms with that paper. The papers [29], on abstract commensurators, and [30], on 'germs of automorphisms' are also relevant. Neretin's groups are also the inspiration for the simple groups acting on trees recently constructed in [31].

Example 5. *An almost automorphism of an infinite, locally finite tree* $\mathcal{T}$ *is a bijection on the vertices of* $\mathcal{T}$ *which preserves all but finitely many edge relations. The set of almost automorphisms forms a group under composition of bijections. This group has two subgroups which are important for this discussion: the group* $\mathrm{Aut}(\mathcal{T})$ *of automorphisms of* $\mathcal{T}$*; and the group* $\mathrm{FSym}(V)$ *of finite permutations of the vertices of* $\mathcal{T}$*. Fix a vertex v in* $\mathcal{T}$ *and let* $U = \mathrm{stab}_{\mathrm{Aut}(\mathcal{T})}(v)$*. Then, U is a compact group under the subspace topology of* $\mathrm{Aut}(\mathcal{T})$ *and each almost automorphism of* $\mathcal{T}$ *commensurates some open subgroup of U to another open subgroup. Hence,*

$$\{xV \mid x \text{ an almost automorphism and } V \text{ an open subgroup of } \mathcal{U}\}$$

is a sub-base of a group topology on the group of almost automorphisms of $\mathcal{T}$. Since $\mathcal{U}$ is open in this topology, the group of almost automorphisms is then a locally compact group. Since non-trivial elements of $FSym(V)$ cannot be in $\mathcal{U}$, the subgroup $FSym(V)$ is closed in this topology and is easily seen to be normal as well. (It may be seen that $FSym(V)$ is the quasi-centre of the group of almost automorphisms, see [32] for the definition.) The quotient of the group of almost automorphisms by $FSym(V)$ is therefore a locally compact group which will be denoted by $AAut(\mathcal{T})$. It is this quotient group which will from now on be referred to as the group of almost automorphisms of $\mathcal{T}$.

Alternative but equivalent definitions of $AAut(\mathcal{T})$ are used elsewhere. For example, in [28], two almost automorphisms of $\mathcal{T}$ are defined to be equivalent if they agree on the complement of some finite subtree of $\mathcal{T}$ and $AAut(\mathcal{T})$ is defined to be the set of equivalence classes of almost automorphisms. Since each finite set of vertices in $\mathcal{T}$ spans a finite subtree, this is the same as the equivalence relation of two almost automorphisms agreeing modulo $FSym(V)$. Almost automorphisms of $\mathcal{T}$ may be seen to be equivalent if and only if the actions they induce on the boundary, $\partial\mathcal{T}$, of the tree agree. The group $AAut(\mathcal{T})$ may thus be defined in terms of its action on $\partial\mathcal{T}$. In these terms, $AAut(\mathcal{T})$ is the full group of the action on $Aut(\mathcal{T})$ on $\partial\mathcal{T}$.

Almost automorphisms of the rooted tree $\mathcal{T}_{q,r}$, in which the root has r children and every other vertex has q children, have been studied extensively. In this notation, Neretin's group of almost automorphisms is $AAut(\mathcal{T}_{2,2})$. The group $AAut(\mathcal{T}_{q,r})$ has the Higman–Thompson group $G_{q,r}$, see [33,34], as a dense subgroup and elements of $G_{q,r}$ may be represented (non-uniquely) as pairs, $(\mathcal{F}_1, \mathcal{F}_2)$, of finite rooted subtrees of $\mathcal{T}_{q,r}$, see [33] or ([35] Section 3) for example. Since the scale is continuous, it therefore suffices, in order to compute the scale on $AAut(\mathcal{T}_{q,r})$, to compute it for elements represented by such pairs of finite trees. This calculation is intricate and the general case is not described in full detail here. Instead, the ideas will be illustrated by the calculation of $s(x)$ for one element x in $G_{2,2} < AAut(\mathcal{T}_{2,2})$. Note, however, that this example only displays some of the intricacies arising in the calculation of the scale on $AAut(\mathcal{T}_{q,r})$.

Let x be the element of $G_{2,2}$ (also known as Thompson's group V) described by the pair of trees $(\mathcal{F}_1, \mathcal{F}_2)$ in Figure 1. When $G_{2,2}$ is embedded in $AAut(\mathcal{T}_{2,2})$, this element denotes the almost automorphism which sends the vertices labelled $1, \ldots, 6$ in the tree on the left to the vertices with the corresponding labels $1', \ldots, 6'$ in the tree on the right and copies across the subtrees below each of these vertices. Thus, the subtree of $\mathcal{F}_1$ whose root is the vertex on level 3 and labelled 1 is raised to a subtree with root on level 2, the subtrees with roots on level 3 and labelled 2 and 3 are copied across to two other subtrees with roots on level 3, the subtrees with roots on level 4 and labelled 4 and 5 are raised to subtrees with roots on level 2 and 3 respectively, and the subtree with root on level 1 and labelled 6 is lowered to a vertex on level 3.

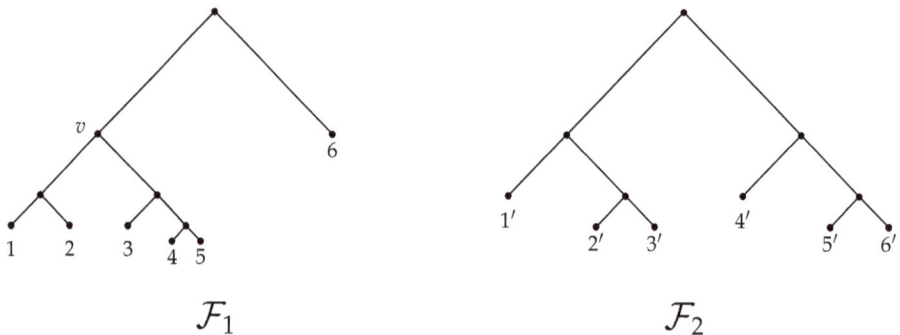

Figure 1. Pair of trees for the element x of Thompson's group V.

Let $U = Aut(\mathcal{T}_{2,2})$, so that U is a compact, open subgroup of $AAut(\mathcal{T}_{2,2})$, and let $\tilde{U}$ be the subgroup of U consisting of all automorphisms which fix the vertices labelled $1, \ldots, 6$ in $\mathcal{F}_1$. Then, $x\tilde{U}x^{-1}$ is the subgroup of U consisting of all automorphisms fixing the vertices labelled $1', \ldots, 6'$ in $\mathcal{F}_2$. Since $\tilde{U}$ and $x\tilde{U}x^{-1}$ are both subgroups of U, it follows that $U \cap x^{-1}Ux \geq \tilde{U}$. The reverse inclusion may be verified by checking cases for automorphisms not in $\tilde{U}$. For example, if $u \in Aut(\mathcal{T}_{2,2})$ interchanges the two vertices on level 1 of the tree, then $x^{-1}ux$ maps the vertices of $\mathcal{F}_1$ labelled 5 and 6, whose only common ancestor is the root, to vertices which have the vertex labelled v as a common ancestor, and no such map is an automorphism of the tree. Hence,

$$[x U x^{-1} : x U x^{-1} \cap U] = [U : \tilde{U}] = 32.$$

That this is not the minimum possible index will be seen by applying the tidying procedure to the subgroup U.

Step 1 It turns out that $U_{-1} = U \cap x^{-1}Ux = \tilde{U}$ is tidy above for x. To see this, Remark 1 tells us that it suffices to show that $\tilde{U}_1 \subseteq \tilde{U}_+\tilde{U}_{-1}$, where $\tilde{U}_1 = \tilde{U} \cap x\tilde{U}x^{-1}$ and $\tilde{U}_{-1} = \tilde{U} \cap x^{-1}\tilde{U}x$. For this, observe that, since $\tilde{U}$ is the fixator of $\mathcal{F}_1$ and $x\tilde{U}x^{-1}$ is the fixator of $\mathcal{F}_2$, $\tilde{U}_1$ is the fixator of $\mathcal{F}_1 \cup \mathcal{F}_2$, see Figure 2. Furthermore, $\tilde{U}_{-1}$ is the fixator of the tree $\mathcal{F}_+$ shown in Figure 3 because x maps the vertices $1^\dagger, \ldots, 5^\dagger$ to the vertices $1, \ldots, 5$; and $\tilde{U}_+$ is the fixator of the tree $\mathcal{V}_+$ shown in Figure 4. ($\mathcal{V}_+$ includes the infinite path spanned by the images of the vertices 4 and 5 under positive powers of x. This is explained further in the next paragraph.) It follows that $\tilde{U}_+\tilde{U}_{-1}$ is the fixator of the tree $\mathcal{F}_+ \cap \mathcal{V}_+$. This intersection is equal to $\mathcal{F}_1$ and so $\tilde{U}_+\tilde{U}_{-1} = \tilde{U}$, which certainly contains $\tilde{U}_1$. Therefore, $\tilde{U}$ is tidy above. To be consistent with the notation of Section 4.1, $\tilde{U}$ will now be denoted by V.

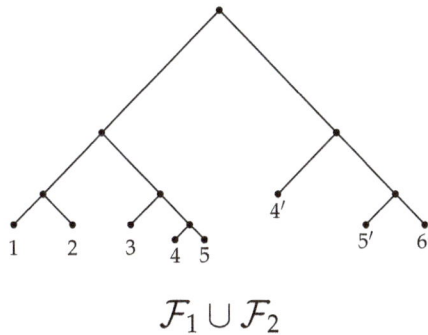

$$\mathcal{F}_1 \cup \mathcal{F}_2$$

Figure 2. $\tilde{U}_1$ is the fixator of the tree $\mathcal{F}_1 \cup \mathcal{F}_2$.

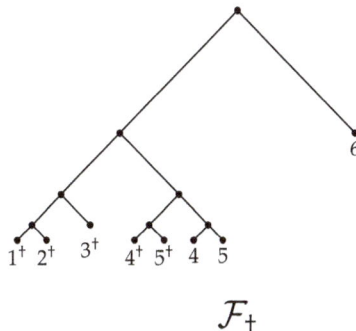

$$\mathcal{F}_\dagger$$

Figure 3. $\tilde{U}_{-1}$ is the fixator of the tree $\mathcal{F}_\dagger$.

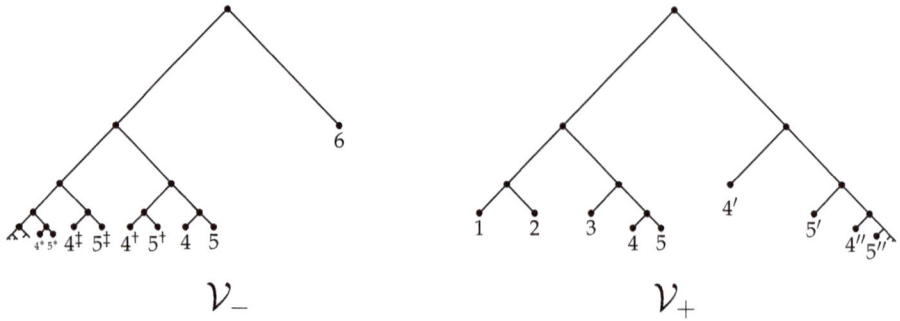

Figure 4. Trees spanned by images of 4 and 5 under powers of x.

Steps 2 and 3 *It further turns out that $\mathcal{L}_V \leq V$, so that V (that is $\tilde{U}$) is tidy below as well. To see this, it suffices to show that, if $v \in V_+$ and $x^N v x^{-N} \in V_-$ for some $N > 0$, then $v \in V_+ \cap V_-$.*

By definition, $V_+ = \bigcap_{n \geq 0} x^n V x^{-n}$. As x is iterated, the vertices 4 and 5 are pushed down two levels of $\mathcal{T}_{2,2}$ at a time and their images are at a distance 1 from a half-line descending from the vertex 6, see Figure 4. This half-line is part of an "axis" for x that is translated down through distance 2 by x. Since V fixes the vertices 4 and 5, the given element $v \in V_+$ fixes the tree V_+ shown in Figure 4.

By definition, $V_- = \bigcap_{n \leq 0} x^n V x^{-n}$. As x is iterated, the vertices 4 and 5 are carried across the tree until they are the children $4^{\ddagger}$ and $5^{\ddagger}$ of vertex 2, and then pushed down one level of $\mathcal{T}_{2,2}$ at a time, their images being at a distance 2 from a half-line descending from the vertex 1, see Figure 4. This half-line is part of an "axis" for x that is translated up through distance 1 by x. Since V fixes the vertices 4 and 5, V_- fixes the tree V_-, shown in Figure 4, which includes this half-line and all vertices within distance 2 of it. For the particular element v, we have that $x^N v x^{-N} \in V_-$ and so v fixes all vertices in V_- below level N. Since v is a tree automorphism, it follows that v fixes all vertices on the half-line descending from vertex 1. However, it does not follow that v fixes all images of vertices 4 and 5 above level N, and that must be shown in order to prove that $v \in V_+ \cap V_-$. To show this, suppose for example that v interchanges the vertices $4^{\dagger}$ and $5^{\dagger}$. Then, $x v x^{-1}$ interchanges 4 and 5, and so $x v x^{-1}$ is in $\mathrm{Aut}(\mathcal{T}_{2,2}) \setminus V$. However, $x^2 v x^{-2}$ interchanges $4'$ and $5'$ and so is not in $\mathrm{Aut}(\mathcal{T}_{2,2})$. Similarly, $x^n v x^{-n}$ does not belong to $\mathrm{Aut}(\mathcal{T}_{2,2})$ for any $n \geq 2$, which contradicts that $x^N v x^{-N} \in V_-$. Hence, v fixes the vertices $4^{\dagger}$ and $5^{\dagger}$. Similar arguments show that v fixes all images of 4 and 5 in V_-, and hence that $v \in V_-$, as claimed.

Since every $v \in V_+$ for which there is $N \geq 1$ with $x^N v x^{-N} \in V_-$ must be in $V_+ \cap V_-$, we have that $\mathcal{L}_V \leq V$ and hence that V is tidy below for x. The scale of x is therefore equal to

$$[xVx^{-1} : xVx^{-1} \cap V] = [x\tilde{U}x^{-1} : x\tilde{U}x^{-1} \cap \tilde{U}] = 4.$$

Just as for automorphism groups of trees, the calculation of the scale of the non-uniscalar $x \in A\mathrm{Aut}(\mathcal{T}_{2,2})$ involves identifying an 'axis' for x. This axis consists of two half-lines with finite trees attached, one of which is translated through distance 2 and determines an attracting end for x, while the other is translated through distance 1 and determines a repelling end. A dynamical description of the action of an almost automorphism may be used in general for the calculation of the scale, although the dynamics can be more complicated as there may be several (and different numbers of) attracting and repelling ends and the almost automorphism may permute some of them. As seen here, the scale depends on more than just the speed with which the axis is translated towards or away from the ends, but also on the "thickness" of the axis. A similar description of the dynamics of the action of almost automorphisms is given in [36], which develops ideas in [37].

5. Computing in t.d.l.c. Groups

In the examples, computing the scale of the element or automorphism of G requires a description of the element or automorphism, a description of a compact open subgroup, U say, of G and a method for calculating the images of U under powers of the automorphism and forming their intersections. The different ways in which these things are done depends on the different concrete representations of G in each case.

It is a truism that computation in a group depends on the description of the group. The computations may be at the level of an abstract group described by a presentation or, for some classes of groups, through a concrete representation. For example, finite groups have concrete representations as permutations (via Cayley's Theorem) and also as matrices (via the regular representation). Lie groups, too, have concrete representations as groups of isometries of symmetric spaces and also as groups of matrices (via the Lie algebra). The two types of concrete representation that exist in the cases of finite and Lie groups might be characterised as *geometric* and *algebraic*. They correspond to the two ways of thinking evident in synthetic and analytic geometry and perhaps too in analogue and digital computing.

Concrete descriptions of t.d.l.c. groups fit the pattern of being geometric or algebraic only to a more limited extent. The automorphism groups of trees in Example 2 are described in geometric terms and the scale is calculated in these terms as well, while the p-adic Lie groups in Example 4 are described and their scale calculated in algebraic terms. Some geometric and algebraic realisations of other t.d.l.c. groups, and the limitations of such represenations, are sketched in the next few paragraphs.

Many t.d.l.c. groups are defined geometrically as automorphism groups of buildings, see [15,38,39], and semisimple Lie groups over local fields. *Kac-Moody groups* over finite fields may also be represented as acting on buildings. Moreover, just as is the case for finite groups and Lie groups, every compactly generated t.d.l.c. group has a geometric representation via an action on a *Cayley–Abels graph*, [40,41]. *Cayley–Abels graphs* are unique only up to quasi-isometry however and, although they can be used to derive bounds on integer invariants of t.d.l.c. groups, see ([30] Proposition 4.6) and [42,43], they do not provide an effective method for performing precise calculations of invariants such as the scale unless the graph structure is understood in as much detail as it is for buildings. The limitation of the *Cayley–Abels geometric* representation therefore is that, unlike the cases of geometries for finite and Lie groups, it is not well understood for general t.d.l.c. groups: no such graph has been described for Neretin's group for example.

The limitation of algebraic realisations (strictly conceived) is that t.d.l.c. groups such as $\mathrm{Aut}(\mathcal{T}_{q+1})$ and Neretin's group do not have finite-dimensional linear representations. However, there is a possible substitute for algebraic realisations of these groups. The calculation of the scale on Neretin's group illustrated in Example 5 uses an approach that is not readily characterised as geometric or algebraic. Although the axes of translation featuring in the dynamical description of the almost automorphism are geometric, the axes and translation distance alone do not determine the scale. The full information required is encoded in the pair of finite trees representing the almost automorphism. This information is combinatorial in nature and says how the almost automorphism commensurates the totally disconnected compact group $\mathrm{Aut}(\mathcal{T}_{2,2})$. Since abstract t.d.l.c. groups always contain a compact open, and hence commensurated, subgroup, they may be realised concretely as groups of commensurators quite generally, see [29,30,35]. Moreover, given any pair (G, H) such that H is a commensurated subgroup of G, a t.d.l.c. group $\widetilde{G}$, called the *relative profinite completion*, may be defined in which the closure of H is a compact open subgroup of $\widetilde{G}$, see [44] and the references therein. Examples of such pairs include the group $PSL(n, \mathbb{Z}[1/p])$ and its commenustared subgroup $PSL(n, \mathbb{Z})$, in which case the relative profinite completion is isomorphic to $PSL(n, \mathbb{Q}_p)$; and the Baumslag–Solitar group $BS(m, n) := \langle a, t : ta^m t^{-1} = a^n \rangle$ and commensurated subgroup $\langle a \rangle$, in which case the relative profinite completion is described in [45].

These examples suggest that a suitable substitute for the adjoint representation might be to represent t.d.l.c. groups concretely as commensurators. This idea is lent support by the fact that, in *p*-adic Lie groups, locally normal subgroups (in the local structure theory developed in [46]) correspond to ideals in the Lie algebra, and that the scale (which is defined in terms of commensuration) may be expressed in terms of eigenvalues. Representing general t.d.l.c. groups as commensurator groups poses challenges comparable with those facing the use of *Cayley–Abels graphs* however.

A practical test of any description of a t.d.l.c. group is whether it facilitates calculation of the scale and identification of tidy subgroups. Seeking descriptions which pass that test for more groups is a task for further investigations. These investigations might guide a possible"classification" of topologically simple t.d.l.c. groups in which the groups are arranged into types according to their best method of concrete description. While linear representations largely suffice in the cases of Lie and finite groups, that will not be the case for t.d.l.c. groups because automorphism groups of trees, Neretin's group and most *Kac–Moody groups* are not linear, and there may be many others yet to be discovered. Geometric, linear and commensurator descriptions may all be required but it is not clear that they will suffice.

S. Smith has shown, see [47], how uncountably many simple discrete groups may be used to produce uncountably many topologically simple, compactly generated t.d.l.c. groups. Smith's groups act on trees (not locally finite ones though) and so have a geometric description, modulo an infinite discrete group, which facilitates calculation of the scale. There seems to be no reason that there might not be other ways of constructing uncountable families of simple t.d.l.c. groups however. These, presumably, would also be described modulo objects in some class known to be uncountable. Arranging groups constructed in these ad hoc ways according to their method of concrete description, and their local structure as defined in [46], may be the best that can be hoped for in the direction of a classification.

Calculating the scale of an element of a t.d.l.c. group can use only a finite amount of information. To calculate the scale, at least in principle, of any element in any compactly generated, topologically simple t.d.l.c. group would, since there are uncountably many of them, entail describing the group up to some finite approximation, or modulo information not relevant to the calculation, and then identifying the element in it up to a finite approximation. It would ultimately be desirable to implement this calculation in computer software. Indeed, it might be argued that such implementation would be the benchmark of success for a theory of t.d.l.c. groups and a categorising, or sorting into types, of the simple ones.

Acknowledgments: The support of the ARC Discovery Grant DP150100060 is gratefully acknowledged.

Conflicts of Interest: The author declares no conflict of interest.

References

1. Van Dantzig, D. Zur topologischen Algebra III: Brouwersche und Cantorsche Gruppen. *Compos. Math.* **1936**, *3*, 408–426.
2. Hewitt, E.; Ross, K.A. *Abstract Harmonic Analysis I*; Grundlehren der mathematischen Wissenschaften, Bd 115; Springer: Berlin/Göttingen/Heidelberg, Germany, 1963.
3. Möller, R.G. Structure theory of totally disconnected locally compact groups via graphs and permutations. *Can. J. Math.* **2002**, *54*, 795–827.
4. Willis, G.A. The structure of totally disconnected, locally compact groups. *Math. Ann.* **1994**, *300*, 341–363.
5. Willis, G.A. Further properties of the scale function on totally disconnected groups. *J. Algebra* **2001**, *237*, 142–164.
6. Willis, G.A. Tidy subgroups for commuting automorphisms of totally disconnected groups: An analogue of simultaneous triangularisation of matrices. *N. Y. J. Math.* **2004**, *10*, 1–35.
7. Jaworski, W.; Rosenblatt, J.M.; Willis, G.A. Concentration functions in locally compact groups. *Math. Ann.* **1996**, *305*, 673–691.
8. Previts, W.H.; Wu, T.-S. On tidy subgroups of locally compact totally disconnected groups. *Bull. Aust. Math. Soc.* **2002**, *65*, 485–490.

9.   Willis, G.A. The scale and tidy subgroups for endomorphisms of totally disconnected locally compact groups. *Math. Ann.* **2015**, *361*, 403–442.
10. Willis, G.A. Totally disconnected groups and proofs of conjectures of Hofmann and Mukherjea. *Bull. Aust. Math. Soc.* **1995**, *51*, 489–494.
11. Caprace, P.-E.; Reid, C.D.; Willis, G.A. Limits of Contraction Groups and the Tits Core. *J. Lie Theory* **2014**, *24*, 957–967.
12. Hofmann, K.H.; Willis, G.A. Continuity characterizing totally disconnected locally compact groups. *J. Lie Theory* **2015**, *25*, 1–7.
13. Braconnier, J. Groupes d'automorphismes d'un groupe localement compact. *Comptes Rendus Acad. Sci. Paris* **1945**, *220*, 382–384.
14. Willis, G.A. Conjugation weights and weighted convolution algebras on totally disconnected, locally compact groups. In *Proceedings of the Centre for Mathematics and Its Applications, Proceedings of the AMSI Conference on Harmonic Analysis and Applications, Macquarie University, Sydney, Australia, 7–11 February 2011*; Duong, X., Hogan, J., Meaney, C., Sikora, A., Eds.; The Australian National University, Mathematical Sciences Institute, Centre for Mathematics & its Applications: Canberra, Australian, 2012; Volume 45, pp. 136–147.
15. Tits, J. Sur le groupe des automorphismes d'un arbre. In *Essays on Topology and Related Topics (Mémoires Dédiés à Georges de Rham)*; Springer: New York, NY, USA, 1970; pp. 188–211.
16. Abramenko, P.; Brown, K.S. *Buildings: Theory and Application*; Graduate Texts in Mathematics; Springer: New York, NY, USA, 2008; Volume 248.
17. Glöckner, H. Scale functions on *p*-adic Lie groups. *Manuscr. Math.* **1998**, *97*, 205–215.
18. Wang, J.S.P. The Mautner phenomenon for p-adic Lie groups. *Math. Z.* **1984**, *185*, 403–412.
19. Baumgartner, U.; Willis, G.A. Contraction groups for automorphisms of totally disconnected groups. *Isr. J. Math.* **2004**, *142*, 221–248.
20. Glöckner, H. Scale Functions on Linear Groups Over Local Skew Fields. *J. Algebra* **1998**, *205*, 525–541.
21. Glöckner, H.; Willis, G.A. Uniscalar *p*-adic Lie groups. *Forum Math.* **2001**, *13*, 413–421.
22. Parreau, A. Sous-groupes elliptiques de groupes linéaires sur un corps valué. (French) [Elliptic subgroups of linear groups over a field with valuation]. *J. Lie Theory* **2003** *13*, 271–278.
23. Kepert, A.; Willis, G.A. Scale functions and tree ends. *J. Aust. Math. Soc.* **2001**, *70*, 273–292.
24. Willis, G.A. Compact open subgroups in simple totally disconnected groups. *J. Algebra* **2007**, *312*, 405–417.
25. Neretin, Y.A. Combinatorial analogues of the group of diffeomorphisms of the circle. (Russian. Russian summary). *Izv. Ross. Akad. Nauk Ser. Mat.* **1992**, *56*, 1072–1085; translation in *Russ. Acad. Sci. Izv. Math.* **1993**, *41*, 337–349.
26. Neretin, Yu.A. Groups of hierarchomorphisms of trees and related hilbert spaces. *J. Funct. Anal.* **2003**, *200*, 505–535.
27. Kapoudjian, C. Simplicity of Neretin's group of spheromorphisms. *Ann. Inst. Fourier (Grenoble)* **1999**, *49*, 1225–1240.
28. Bader, U.; Caprace, P.-E.; Gelander, T.; Mozes, S. Simple groups without lattices. *Bull. Lond. Math. Soc.* **2012**, *44*, 55–67.
29. Barnea, Y.; Ershov, M.; Weigel, T. Abstract commensurators of profinite groups. *Trans. Am. Math. Soc.* **2011**, *363*, 5381–5417.
30. Caprace, P.-E.; De Medts, T. Simple locally compact groups acting on trees and their germs of automorphisms. *Transform. Groups* **2011**, *16*, 375–411.
31. Le Boudec, A. Groups acting on trees with almost prescribed local action. *Comment. Math. Helv.* **2016**, *91*, 253–293.
32. Burger, M.; Mozes, S. Groups acting on trees: From local to global structure. *Inst. Hautes Études Sci. Publ. Math.* **2000**, *92*, 113–150.
33. Cannon, J.W.; Floyd, W.J.; Parry, W.R. Introductory notes on Richard Thompson's groups. *Enseign. Math.* **1996**, *42*, 215–256.
34. Higman, G. Finitely presented infinite simple groups. In *Notes on Pure Mathematics 8*; Australian National University: Canberra, Australia, 1974.
35. Röver, C. Abstract commensurators of groups acting on rooted trees. *Geom. Dedicata* **2002**, *94*, 45–61.
36. Salazar-Díaz, O. P. Thompson's group *V* from a dynamical viewpoint. *Int. J. Algebra Comput.* **2010**, *20*, 39–70.
37. Brin, M. Higher dimensional Thompson's groups. *Geom. Dedicata* **2004**, *108*, 163–192.

38. Haglund, F.; Paulin, F. Simplicité de groupes d'automorphismes d'espaces à courbure négative. *Geom. Topol. Monogr.* **1998**, *1*, 181–248.

39. Nebbia, C. Minimally almost periodic totally disconnected groups. *Proc. Am. Math. Soc.* **1999**, *128*, 347–351.

40. Abels, H. Specker-Kompaktifizierungen von lokal kompakten topologischen Gruppen. *Math. Z.* **1974**, *138*, 325–361.

41. Monod, N. *Continuous Bounded Cohomology of Locally Compact Groups*; Lecture Notes in Mathematics; Springer: Berlin, Germany, **2001**; Volume 1758.

42. Baumgartner, U.; Möller, R.; Willis, G.A. Hyperbolic groups have flat-rank at most 1. *Isr. J. Math.* **2012**, *190*, 365–388.

43. Reid, C.D.; Wesolek, P.R. The essentially chief series of a compactly generated locally compact group. *Math. Ann.* **2017**, doi:10.1007/s00208-017-1597-0.

44. Reid, C.D.; Wesolek, P.R. Homomorphisms in totally disconnected, locally compact groups with dense image. *arXiv* **2015**, doi:arXiv:1509.00156v1.

45. Elder, M.; Willis, G.A. Totally disconnected groups from Baumslag-Solitar groups. *arXiv* **2013**, arXiv:1301.4775.

46. Caprace, P.-E.; Reid, C.D.; Willis, G.A. Locally normal subgroups of totally disconnected groups. Part II: Compactly generated simple groups. *Forum Math. Sigma* **2017**, *5*, doi:10.1017/fms.2017.9.

47. Smith, S.M. A product for permutation groups and topological groups. *Duke Math. J.* **2017**, *166*, 2965–2999.

![axioms logo] *axioms*

MDPI

Article

Extending Characters of Fixed Point Algebras

Stefan Wagner [ORCID]

Department of Mathematics and Natural Sciences, Blekinge Tekniska Högskola, 371 41 Karlskrona, Sweden; stefan.wagner@bth.se

Received: 13 October 2018; Accepted: 5 November 2018; Published: 7 November 2018

Abstract: A dynamical system is a triple (A, G, α) consisting of a unital locally convex algebra A, a topological group G, and a group homomorphism $\alpha : G \to \mathrm{Aut}(A)$ that induces a continuous action of G on A. Furthermore, a unital locally convex algebra A is called a continuous inverse algebra, or CIA for short, if its group of units $A^{\times}$ is open in A and the inversion map $\iota : A^{\times} \to A^{\times}, a \mapsto a^{-1}$ is continuous at 1_A. Given a dynamical system (A, G, α) with a complete commutative CIA A and a compact group G, we show that each character of the corresponding fixed point algebra can be extended to a character of A.

Keywords: dynamical system; continuous inverse algebra; character; maximal ideal; fixed point algebra; extension

MSC: 46H05; 46H10 (primary); 37B05 (secondary)

---

1. Introduction

Let $\sigma : P \times G \to P$ be a smooth action of a Lie group G on a manifold P. It is well-known (see e.g., [1], Proposition 2.1) that σ induces a smooth action of G on the unital Fréchet algebra $C^{\infty}(P)$ of smooth functions on P defined by $\alpha_{\sigma} : G \times C^{\infty}(P) \to C^{\infty}(P), (g, f) \mapsto f \circ \sigma_g$. The corresponding fixed point algebra is given by

$$C^{\infty}(P)^G := \{f \in C^{\infty}(P) : (\forall g \in G)\ \alpha_{\sigma}(g, f) = f\}.$$

The origin of this short article is, in a manner of speaking, "commutative geometry", namely the question whether *each character $\chi : C^{\infty}(P)^G \to \mathbb{C}$ extends to a character $\tilde{\chi} : C^{\infty}(P) \to \mathbb{C}$* (cf. [2,3]).

One possible way to approach this problem is to classify the characters under consideration. Indeed, it follows from ([1], Lemma A.1) that each character $\chi : C^{\infty}(P) \to \mathbb{C}$ is an evaluation in some point $p \in P$, that is, of the form $\delta_p : C^{\infty}(P) \to \mathbb{C}, f \mapsto f(p)$. If the action σ is additionally free and proper, then the orbit space P/G has a unique manifold structure such that the canonical quotient map $q : P \to P/G, p \mapsto [p]$ is a submersion. Moreover, in this situation, the map

$$\Phi : C^{\infty}(P)^G \to C^{\infty}(P/G), \quad f \mapsto ([p] \mapsto f(p))$$

is an isomorphism of unital Fréchet algebras showing that each character $C^{\infty}(P)^G \to \mathbb{C}$ is of the form $\delta_{[p]} \circ \Phi$ for some $p \in P$ which may simply be extended by δ_p.

In this note, however, we approach the above problem in a more systematic way. In fact, given a dynamical system (A, G, α) with a complete commutative continuous inverse algebra (CIA) A and a compact group G, we show that each character of the corresponding fixed point algebra

$$A^G := \{a \in A : (\forall g \in G)\ \alpha(g)(a) = a\}$$

extends to a character of A (Theorem 2). Our approach is motivated by the following three facts:

(i) Our initial question is, after all, of purely topological nature.
(ii) If P is compact, then $C^\infty(P)$ is the prototype of a complete commutative CIA.
(iii) CIA's provide a class of algebras for which characters are automatically continuous (cf. [4], Lemma 2.3).

We would also like to mention that CIAs are naturally encountered in K-theory and noncommutative geometry, usually as dense unital subalgebras of C*-algebras. Finally, we point out that a classical result for actions of finite groups can be found in ([5], Chapter 5, §2.1, Corollary 4).

2. Preliminaries and Notations

All algebras are assumed to be complex. The spectrum of an algebra A is the set $\Gamma_A := \mathrm{Hom}_{\mathrm{alg}}(A, \mathbb{C}) \backslash \{0\}$ (endowed with the topology of pointwise convergence on A) and its elements are called characters. Moreover, given a compact group G, we denote by $\hat{G}$ the (countable) set of equivalence classes of finite-dimensional irreducible representations of G. For $\pi \in \hat{G}$ we write χ_π for the function defined by $G \mapsto \mathbb{C}$, $g \mapsto \mathrm{tr}(\pi(g))$ and we put $d_\pi := \chi_\pi(1_G)$ for the corresponding dimension. We also need the following well-known structure theorem for dynamical systems:

Lemma 1. ([6], [Lemma 3.2 and Theorem 4.22]). *Let (A, G, α) be a dynamical system with a complete unital locally convex algebra A and a compact group G. Furthermore, given $\pi \in \hat{G}$ and $a \in A$, let*

$$P_\pi(a) := d_\pi \int_G \overline{\chi_\pi}(g)\, (\alpha(g)(a))\, dg,$$

where dg denotes the normalized Haar measure on G. Then the following assertions hold:

(a) *For each $\pi \in \hat{G}$ the map $P_\pi : A \to A$ is a continuous G-equivariant projection onto the G-invariant subspace $A_\pi := P_\pi(A)$. In particular, A_π is algebraically and topologically a direct summand of A.*
(b) *The module direct sum $A_{\mathrm{fin}} := \bigoplus_{\pi \in \hat{G}} A_\pi$ is a dense subalgebra of A.*

3. Extension Results

In this section our main results are stated and proved. We begin with some general statements on the extendability of ideals.

Lemma 2. *Let (A, G, α) be a dynamical system with a complete unital locally convex algebra A and a compact group G. Then the following assertions hold:*

(a) *If I is a proper left ideal in A^G, then $A_{\mathrm{fin}} \cdot I = \bigoplus_{\pi \in \hat{G}} A_\pi \cdot I$ defines a proper left ideal in A_{fin} that contains I.*
(b) *If I is a proper closed left ideal in A^G and J is the closure of $A_{\mathrm{fin}} \cdot I$ in A_{fin}, then J is a proper closed left ideal in A_{fin} that contains I.*

Proof. (a) We first observe that A^G coincides with A_1 (where 1 stands for the equivalence class of the trivial representation). Hence $I \subseteq A^G$ is contained in A_{fin} and thus $A_{\mathrm{fin}} \cdot I$ is the left ideal of A_{fin} generated by I. Using the integral formula for P_π from Lemma 1, we see that $A_\pi \cdot I \subseteq A_\pi$, entailing that the sum in part (a) is direct. To see that $A_{\mathrm{fin}} \cdot I$ is proper, we assume the contrary, that is,

$$1_A \in A_{\mathrm{fin}} \cdot I = \bigoplus_{\pi \in \hat{G}} A_\pi \cdot I.$$

Then $1_A \in A^G$ implies that $1_A \in A^G \cdot I = I$, which contradicts the fact that I is a proper left ideal of A^G. We conclude that $A_{\mathrm{fin}} \cdot I$ is a proper left ideal in A_{fin} that contains I.

(b) Part (a) and the definition of J imply that J is a closed left ideal in A_{fin} that contains I. To see that J is proper, we again assume the contrary, that is, $1_A \in J$. Then there exists a net $(a_\gamma)_{\gamma \in \Gamma}$ in $A_{\mathrm{fin}} \cdot I$

such that $\lim_\gamma a_\gamma = 1_A$ and the continuity of the projection map $P_1 : A \to A$ onto the fixed point algebra A^G implies that

$$1_A = P_1(1_A) = P_1(\lim_\gamma a_\gamma) = \lim_\gamma P_1(a_\gamma).$$

Since I is closed in A^G and $P_1(a_\gamma) \in A^G \cdot I = I$ for all $\gamma \in \Gamma$, we conclude that $1_A \in I$. This contradicts the fact that I is a proper ideal of A^G and therefore J is a proper closed left ideal in A_{fin} that contains I. $\square$

Lemma 3. *Let A be a topological algebra and B a dense subalgebra of A. If I is a proper closed left ideal in B, then $\bar{I}$ is a proper closed left ideal in $\bar{B} = A$.*

Proof. It is easily seen that $\bar{I}$ is a closed left ideal in $\bar{B} = A$. Moreover, we have $I = \bar{I} \cap B$. Indeed, the inclusion " $\subseteq$ " is obvious and for the other inclusion we use the fact that I is closed in B. Consequently, if $\bar{I}$ is not proper, that is, $\bar{I} = A$, then $I = B$, which yields a contradiction. Hence, $\bar{I}$ is a proper closed left ideal in A. $\square$

We are now ready to state and prove our main extension results.

Theorem 1. (*Extending ideals*). *Let (A, G, α) be a dynamical system with a complete unital locally convex algebra A and a compact group G. Then each proper closed left ideal in A^G is contained in a proper closed left ideal in A.*

Proof. Let I be a proper closed left ideal in A^G. Then Lemma 2 (b) implies that I is contained in a proper closed left ideal in A_{fin}. Since A_{fin} is a dense subalgebra of A by Lemma 1 (b), the claim is a consequence of Lemma 3. $\square$

Theorem 2. (*Extending characters*). *Let (A, G, α) be a dynamical system with a complete commutative CIA A and a compact group G. Then each character $\chi : A^G \to \mathbb{C}$ is continuous and extends to a continuous character $\tilde{\chi} : A \to \mathbb{C}$.*

Proof. Let $\chi : A^G \to \mathbb{C}$ be a character. Since A^G carries the structure of a CIA in its own right, it follows from ([4], Lemma 2.3) that χ is continuous which shows that $I := \ker \chi$ is a proper closed ideal in A^G. Hence, Theorem 1 implies that I is contained in a proper closed ideal in A. In particular, it is contained in a proper maximal ideal J of A which, according to ([7], Lemma 2.2.2) and ([4], Lemma 2.3), is the kernel of some continuous character $\tilde{\chi} : A \to \mathbb{C}$. Since I is a maximal ideal in the unital algebra A^G and

$$I = I \cap A^G \subseteq J \cap A^G \subseteq A^G,$$

we conclude that $I = J \cap A^G$. Therefore, the decomposition $A^G = I \oplus \mathbb{C} = (J \cap A^G) \oplus \mathbb{C}$ finally proves that $\tilde{\chi}$ extends χ. $\square$

Remark 1. *It is not clear how to extend Theorem 2 beyond the class of CIAs. For instance, given a non-compact manifold P, the set $C_c^\infty(P)$ of compactly supported smooth functions on P is a proper ideal in $C^\infty(P)$. As such it is contained in a proper maximal ideal in $C^\infty(P)$ that cannot be closed since $C_c^\infty(P)$ is dense in $C^\infty(p)$. However, in the more general situation of a complete commutative unital locally convex algebra A, a similar argument as in the proof of Theorem 2 shows that each continuous character $\chi : A^G \to \mathbb{C}$ can be extended to a character $\tilde{\chi} : A \to \mathbb{C}$.*

We conclude with the following two immediate corollaries.

Corollary 1. *Suppose we are in the situation of Theorem* 2. *Then the natural map on the level of spectra* $\Gamma_A \to \Gamma_{A^G}$, $\chi \mapsto \chi_{|A^G}$ *is surjective.*

Corollary 2. *Let* $(C^\infty(P), G, \alpha)$ *be a dynamical system with a compact manifold P and a compact group G. Then each character* $\chi : C^\infty(P)^G \to \mathbb{C}$ *extends to a character* $\tilde{\chi} : C^\infty(P) \to \mathbb{C}$.

Remark 2. *Given a dynamical system* $(C^\infty(P), G, \alpha)$ *with a compact manifold P and a compact group G, we would like to describe* $\Gamma_{C^\infty(P)^G}$ *as a set of points associated to P and G. As already explained in the introduction, it is not hard to see that* $\Gamma_{C^\infty(P)^G}$ *is homeomorphic to P/G if G is a Lie group and* α *is induced by a free and smooth action of G on P. However, even if we do not have any additional information, it is still possible to show that the map*

$$P/G \to \Gamma_{C^\infty(P)^G}, \quad q(p) \mapsto \delta_p$$

is a homeomorphism (see e.g., [2]*, Proposition 8.7) and Corollary* 2 *may be used to verify its surjectivity.*

Funding: This research received no external funding.

Acknowledgments: The author thanks Henrik Seppänen and Erhard Neher for useful discussions on this topic. He would also like to express his gratitude to the referees for providing very fruitful criticism that helped to improve the article.

Conflicts of Interest: The author declares no conflict of interest.

References

1.   Wagner, S. Free group actions from the viewpoint of dynamical systems. *Münster J. Math.* **2012**, *5*, 73–97.
2.   Wagner, S. A Geometric Approach to Noncommutative Principal Torus Bundles. *Proc. Lond. Math. Soc.* **2013**, *106*, 1179–1222. [CrossRef]
3.   Wagner, S. On noncommutative principal bundles with finite abelian structure group. *J. Noncommut. Geom.* **2014**, *8*, 987–1022. [CrossRef]
4.   Biller, H. Continuous Inverse Algebras with Involution. *Forum Math.* **2010**, *22*, 1033–1059. [CrossRef]
5.   Bourbaki, N. *Commutative Algebra: Chapters 1–7; Elements of Mathematics (Berlin)*; Translated from the French, Reprint of the 1972 Edition; Springer: Berlin, Germany, 1989; p. xxiv+625.
6.   Hofmann, K.H.; Morris, S.A. *The Structure of Compact Groups*; De Gruyter Studies in Mathematics; A Primer for the Student—A Handbook for the Expert, Third Edition, Revised and Augmented; De Gruyter: Berlin, Germany, 2013; Volume 25, p. xxii+924. [CrossRef]
7.   Biller, H. Continuous Inverse Algebras and Infinite-Dimensional Linear Lie Groups. Ph.D. Thesis, Technische Universität Darmstadt, Darmstadt, Germany, 2004.

axioms

MDPI

Article

A Note on the Topological Group c_0

Michael Megrelishvili

Department of Mathematics, Bar-Ilan University, 52900 Ramat-Gan, Israel; megereli@math.biu.ac.il

Received: 28 September 2018; Accepted: 24 October 2018; Published: 29 October 2018

Abstract: A well-known result of Ferri and Galindo asserts that the topological group c_0 is not reflexively representable and the algebra $\mathrm{WAP}(c_0)$ of weakly almost periodic functions does not separate points and closed subsets. However, it is unknown if the same remains true for a larger important algebra $\mathrm{Tame}(c_0)$ of tame functions. Respectively, it is an open question if c_0 is representable on a Rosenthal Banach space. In the present work we show that $\mathrm{Tame}(c_0)$ is small in a sense that the unit sphere S and $2S$ cannot be separated by a tame function $f \in \mathrm{Tame}(c_0)$. As an application we show that the Gromov's compactification of c_0 is not a semigroup compactification. We discuss some questions.

Keywords: Gromov's compactification; group representation; matrix coefficient; semigroup compactification; tame function

1. Introduction

Recall that for every Hausdorff topological group G the algebra $\mathrm{WAP}(G)$ of all weakly almost periodic functions on G determines the universal semitopological semigroup compactification $u_w : G \to G^w$ of G. This map is a topological embedding for many groups including the locally compact case. For some basic material about $\mathrm{WAP}(G)$ we refer to [1,2].

The question if u_w always is a topological embedding (i.e., if $\mathrm{WAP}(G)$ determines the topology of G) was raised by Ruppert [2]. This question was negatively answered in [1] by showing that the Polish topological group $G := H_+[0,1]$ of orientation preserving homeomorphisms of the closed unit interval has only constant WAP functions and that every continuous representation $h : G \to Is(V)$ (by linear isometries) on a reflexive Banach space V is trivial. The WAP triviality of $H_+[0,1]$ was conjectured by Pestov.

Recall also that for $G := H_+[0,1]$ every Asplund (hence also every WAP) function is constant and every continuous representation $G \to Iso(V)$ on an Asplund (hence also reflexive) space V must be trivial [3]. In contrast one may show (see [4,5]) that $H_+[0,1]$ is representable on a (separable) Rosenthal space (a Banach space is *Rosenthal* if it does not contain a subspace topologically isomorphic to l_1).

We have the inclusions of topological G-algebras

$$\mathrm{WAP}(G) \subset \mathrm{Asp}(G) \subset \mathrm{Tame}(G) \subset \mathrm{RUC}(G).$$

For details about $\mathrm{Tame}(G)$ and definition of $\mathrm{Asp}(G)$ see [5–7]. We only remark that $f \in \mathrm{Tame}(G)$ if and only if f is a matrix coefficient of a Rosenthal representation. That is, there exist: a Rosenthal Banach space V; a continuous homomorphism $h : G \to Is(V)$ into the topological group of all linear isometries $V \to V$ with strong operator topology; two vectors $v \in V$; $\psi \in V^*$ (the dual of V) such that $f(g) = \psi(h(g)v)$ for every $g \in G$.

Similarly, it can be characterized $f \in \mathrm{Asp}(G)$ replacing Rosenthal spaces by the larger class of Asplund spaces. A Banach space is *Asplund* if the dual of every separable subspace is separable. Every reflexive space is Asplund and every Asplund is Rosenthal. A standard example of an Asplund but nonreflexive space is just c_0.

Recall that c_0, as an additive abelian topological group, is not representable on a reflexive Banach space by a well-known result of Ferri and Galindo [8]. In fact, $WAP(c_0)$ separates the points but not points and closed subsets. The group c_0 admits an injective continuous homomorphism $h : c_0 \to Is(V)$ with some reflexive V but such h cannot be a topological embedding.

Presently it is an open question if every topological group (abelian, or not) G is Rosenthal representable and if $\text{Tame}(G)$ determines the topology of G. Note that the algebra $\text{Tame}(G)$ appears as an important modern tool in some new research lines in topological dynamics motivating its detailed study [5,7].

One of the good reasons to study $\text{Tame}(G)$ is a special role of tameness in the dynamical Berglund-Fremlin-Talagrand dichotomy [5]; as well as direct links to Rosenthal's l_1-dychotomy. In a sense $\text{Tame}(G)$ is a set of all functions which are not dynamically massive.

By these reasons and since $H_+[0, 1]$ is Rosenthal representable, it seems to be an attractive concrete question if c_0 is Rosenthal representable and it is worth studying how large is $\text{Tame}(c_0)$. In the present work we show that $\text{Tame}(c_0)$ is quite small (even for the discrete copy of c_0, see Theorem 3).

Theorem 1. *Tame(c_0) does not separate the unit sphere S and $2S$.*

So, the closures of S and $2S$ intersect in the universal tame compactification of c_0 (a fortiori, the same is true for the universal Asplund (HNS) semigroup compactification).

Another interesting question is if c_0 admits an embedding into a *metrizable* semigroup compactification. Note that any metrizable semigroup compactification of $H_+[0, 1]$ is trivial.

In Section 3 we show that the Gromov's compactification $\gamma : c_0 \hookrightarrow P$, which is metrizable (and γ is a G-embedding), is not a semigroup compactification.

Theorem 2. *Let $\gamma : c_0 \hookrightarrow P$ be the Gromov's compactification of the metric space $(c_0, \frac{d}{1+d})$, where $d(x, y) := ||x - y||$. Then γ is not a semigroup compactification.*

This gives an example of a naturally defined separable unital (original topology determining) G-subalgebra of $\text{RUC}(G)$ (for $G = c_0$) which is not left m-introverted in the sense of [9].

2. Tame Functions on c_0

Recall that a sequence f_n of real-valued functions on a set X is said to be *independent* if there exist real numbers $a < b$ such that

$$\bigcap_{n \in P} f_n^{-1}(-\infty, a) \cap \bigcap_{n \in M} f_n^{-1}(b, \infty) \neq \varnothing$$

for all finite disjoint subsets P, M of $\mathbb{N}$. Every bounded independent sequence is an l_1-sequence [10].

As in [6,7] we say that a bounded family F of real-valued (not necessarily continuous) functions on a set X is a *tame family* if F does not contain an independent sequence.

Let G be a topological group, $f : G \to \mathbb{R}$ be a real-valued function. For every $g \in G$ define $fg : G \to \mathbb{R}$ as $(fg)(x) = f(gx)$ (for multiplicative G). Denote by $\text{RUC}(G)$ the algebra of all bounded right uniformly continuous functions on G. So, $f \in \text{RUC}(G)$ means that f is bounded and for every $\epsilon > 0$ there exists a neighborhood U of the identity e (of the multiplicative group G) such that $|f(ux) - f(x)| < \epsilon$ for every $x \in G$ and $u \in U$. This algebra $\text{RUC}(G)$ corresponds to the greatest G-compactification $G \to \beta_G G$ of G (with respect to the left action), *greatest ambit* of G.

We say that $f \in \text{RUC}(G)$ is a tame function if the orbit $fG := \{fg\}_{g \in G}$ is a tame family. That is, fG does not contain an independent sequence; notation $f \in \text{Tame}(G)$.

2.1. Proof of Theorem 1

We have to show that $\text{Tame}(c_0)$ does not separate the spheres S and $2S$ (where $S := \{x \in c_0 : \|x\| = 1\}$). In fact we show the following stronger result.

Theorem 3. *Let $G = c_0$ be the additive group of the classical Banach space c_0. Assume that $f : c_0 \to \mathbb{R}$ be any (not necessarily continuous) bounded function such that*

$$\begin{cases} f(x) \le a & \forall \, \|x\| = 1 \\ b \le f(x) & \forall \, \|x\| = 2 \end{cases}$$

for some pair $a < b$ of real numbers. Then f is not a tame function on the discrete copy of the group c_0.

Proof. For every $n \in \mathbb{N}$ consider the function

$$f_n : c_0 \to \mathbb{R}, x \mapsto f(e_n + x),$$

where e_n is a vector of c_0 having 1 as its n-th coordinate and all other coordinates are 0. Clearly, $f_n = fg_n$ where $g_n = e_n \in c_0$. We have to check that fG is an untame family. It is enough to show that the sequence $\{f_n\}_{n \in \mathbb{N}}$ in fG is an independent family of functions on c_0. We have to show that for every finite nonempty disjoint subsets I, J in $\mathbb{N}$ the intersection

$$\bigcap_{n \in I} f_n^{-1}(-\infty, a] \cap \bigcap_{n \in J} f_n^{-1}[b, \infty)$$

is nonempty.

Define $v = (v_k)_{k \in \mathbb{N}} \in c_0$ as follows: $v_j = 1$ for every $j \in J$ and $v_k = 0$ for every $k \notin J$. Then

(1) $v \in c_0$ and $\|v\| = 1$.
(2) $\|e_i + v\| = 1$, $f_i(v) = f(e_i + v) \le a$ for every $i \in I$.
(3) $\|e_j + v\| = 2$, $f_j(v) = f(e_j + v) \ge b$ for every $j \in J$.

So we found v such that

$$v \in \bigcap_{n \in I} f_n^{-1}(-\infty, a] \cap \bigcap_{n \in J} f_n^{-1}[b, \infty).$$

□

Corollary 1. *The bounded RUC function*

$$f : c_0 \to [-1, 1], x \mapsto \frac{\|x\|}{1 + \|x\|}$$

is not tame on c_0 (even on the discrete copy of the group c_0).

Proof. Observe that $f(S) = \frac{1}{2}, f(2S) = \frac{2}{3}$ and apply Theorem 3. □

Theorem 3 remains true for the spheres rS and $2rS$ for every $r > 0$. In the case of Polish c_0 it is unclear if the same is true for any pair of different spheres around the zero. If, yes then this will imply that $\text{Tame}(c_0)$ does not separate the zero and closed subsets. The following question remains open even for any topological group [5,7].

Question 1. *Is it true that $\text{Tame}(c_0)$ separates the points and closed subsets ? Is it true that Polish group c_0 is Rosenthal representable ?*

3. Gromov's Compactification Need Not Be a Semigroup Compactification

Studying topological groups G and their dynamics we need to deal with various natural closed unital G-subalgebras $\mathcal{A}$ of the algebra $\mathrm{RUC}(G)$. Such subalgebras lead to G-compactifications of G (so-called G-*ambits*, [11]). That is we have compact G-spaces K with a dense orbit $Gz \subset K$ such that the Gelfand algebra which corresponds to the compactification $G \to K, g \mapsto gz$ is just $\mathcal{A}$. Frequently but not always such compactifications are the so-called *semigroup compactifications*, which are very useful in topological dynamics and analysis. Compactifications of topological groups already is a fruitful research line. See among others [12–14] and references there. In our opinion semigroup compactifications deserve even much more attention and systematic study in the context of general topological group theory.

A semigroup compactification of G is a pair (α, K) such that K is a compact *right topological semigroup* (all right translations are continuous), and α is a continuous semigroup homomorphism from G into K, where $\alpha(G)$ is dense in K and the left translation $K \to K, x \mapsto \alpha(g)x$ is continuous for every $g \in G$.

One of the most useful references about semigroup compactifications is a book of Berglund, Junghenn and Milnes [9]. For some new directions (regarding topological groups) see also [3,4,15,16].

Question 2. *Which natural compactifications of topological groups G are semigroup compactifications? Equivalently which Banach unital G-subalgebras of $\mathrm{RUC}(G)$ are left m-introverted (in the sense of [9])?*

Recall that *left m-introversion* of a subalgebra $\mathcal{A}$ of $\mathrm{RUC}(G)$ means that for every $v \in \mathcal{A}$ and every $\psi \in MM(\mathcal{A})$ the matrix coefficient $m(v, \psi)$ belongs to $\mathcal{A}$, where

$$m(v, \psi) : G \to \mathbb{R}, g \mapsto \psi(g^{-1}v)$$

and $MM(\mathcal{A}) \subset \mathcal{A}^*$ denotes the spectrum (Gelfand space) of $\mathcal{A}$.

It is not always easy to verify left m-introversion directly. Many natural G-compactifications of G are semigroup compactifications. For example, it is true for the compactifications defined by the algebras $\mathrm{RUC}(G)$, $\mathrm{Tame}(G)$, $\mathrm{Asp}(G)$, $\mathrm{WAP}(G)$. Of course, the 1-point compactification is a semitopological semigroup compactification for any locally compact group G.

As to the counterexamples. As it was proved in [3], the subalgebra $\mathrm{UC}(G) := \mathrm{RUC}(G) \cap \mathrm{LUC}(G)$ of all uniformly continuous functions is not left m-introverted for $G := H(C)$, the Polish group of homeomorphisms of the Cantor set.

In this section we show that the Gromov's compactification of a metrizable topological group G need not be a semigroup compactification.

Let ρ be a bounded metric on a set X. Then the Gromov's compactification of the metric space (X, ρ) is a compactification $\gamma : X \to P$ induced by the algebra $\mathcal{A}$ which is generated by the bounded set of functions

$$\{\rho_z : X \to \mathbb{R}, \rho_z(x) = \rho(z, x)\}_{z \in X}.$$

Then γ always is a topological embedding. If X is separable then P is metrizable. Moreover, if (X, ρ) admits a continuous ρ-invariant action of a topological group G then γ is a G-compactification of X; see [17].

Here we examine the following particular case. Let G be a metrizable topological group. Choose any left invariant metric d on G. Denote by $\gamma : G \to P$ the Gromov's compactification of the bounded metric space (G, ρ), where $\rho = \frac{d}{1+d}$.

Consider the following natural bounded RUC function

$$f : G \to \mathbb{R}, x \mapsto \frac{||x||}{1 + ||x||}$$

where $||x|| := d(e, x)$. By $\mathcal{A}_f$ we denote the smallest closed unital G-subalgebra of $RUC(G)$ which contains $fG = \{fg : g \in G\}$. Then $\mathcal{A}_f$ is the algebra which corresponds to the compactification γ. Indeed, $\rho_{g^{-1}}(x) = \rho(g^{-1}, x) = (fg)(x)$ for every $g, x \in G$.

Proof of Theorem 2

We have to prove Theorem 2.

Proof. By the discussion above, the unital G-subalgebra $\mathcal{A}_f$ of $RUC(G)$ associated with γ is generated by the orbit fG, where $f : G \to \mathbb{R}, f(x) = \frac{||x||}{1+||x||}$. Since c_0 is separable the algebra $\mathcal{A}_f$ is separable. Hence, P is metrizable. If we assume that γ is a semigroup compactification then the separability of $\mathcal{A}_f$ guarantees by [4] (Prop. 6.13) that $\mathcal{A}_f \subset Asp(G)$. On the other hand, since $Asp(G) \subset Tame(G)$, and $f \in \mathcal{A}_f$ we have $f \in Tame(G)$. Now observe that f separates the spheres S and $2S$ and we get a contradiction to Corollary 1. $\square$

Question 3. *Is it true that the Polish group c_0 admits a semigroup compactification $\alpha : c_0 \hookrightarrow P$ such that P is metrizable and α is an embedding? What if P is first countable?*

This question is closely related to the setting of this work. Indeed, by [4] (Prop. 6.13) (resp., by [4] (Cor. 6.20)) the metrizability (first countability) of P guarantees that the corresponding algebra is a subset of $Asp(G)$ (resp. of $Tame(G)$).

Funding: This research received no external funding.

Conflicts of Interest: The author declares no conflicts of interest.

References

1. Megrelishvili, M. Every semitopological semigroup compactification of the group $H_+[0,1]$ is trivial. *Semigroup Forum* **2001**, *63*, 357–370. [CrossRef]
2. W. Ruppert, *Compact Semitopological Semigroups: An Intrinsic Theory*; Lecture Notes in Mathematics, 1079; Springer: New York, NY, USA, 1984.
3. Glasner, E.; Megrelishvili, M. New algebras of functions on topological groups arising from G-spaces. *Fundamenta Math.* **2008**, *201*, 1–51. [CrossRef]
4. Glasner, E.; Megrelishvili, M. Banach representations and affine compactifications of dynamical systems. In *Fields Institute Proceedings Dedicated to the 2010 Thematic Program on Asymptotic Geometric Analysis*; Ludwig, M., Milman, V.D., Pestov, V., Tomczak-Jaegermann, N., Eds.; Springer: New York, NY, USA, 2013.
5. Glasner, E.; Megrelishvili, M. Representations of dynamical systems on Banach spaces. In *Recent Progress in General Topology III*; Hart, K.P., van Mill, J., Simon, P., Eds.; Atlantis Press: Amsterdam, The Netherlands, 2014; pp. 399–470.
6. Glasner, E.; Megrelishvili, M. Representations of dynamical systems on Banach spaces not containing l_1. *Trans. Am. Math. Soc.* **2012**, *364*, 6395–6424. [CrossRef]
7. Glasner, E.; Megrelishvili, M. More on tame dynamical systems. In *Lecture Notes S. vol. 2013, Ergodic Theory and Dynamical Systems in Their Interactions with Arithmetics and Combinatorics*; Ferenczi, S., Kulaga-Przymus, J., Lemanczyk, M., Eds.; Springer: New York, NY, USA, 2018.
8. Ferri, S.; Galindo, J. Embedding a topological group into its WAP-compactification. *Studia Math.* **2009**, *193*, 99–108. [CrossRef]
9. Berglund, J.F.; Junghenn, H.D.; Milnes, P. *Analysis on Semigroups*; Wiley: New York, NY, USA, 1989.
10. Rosenthal, H.P. A characterization of Banach spaces containing ℓ_1. *Proc. Natl. Acad. Sci. USA* **1974**, *71*, 2411–2413. [CrossRef] [PubMed]
11. de Vries, J. *Elements of Topological Dynamics*; Kluwer Academic Publishers: Norwell, MA, USA, 1993.
12. Uspenskij, V.V. Compactifications of topological groups. In Proceedings of the Ninth Prague Topological Symposium, Prague, Czech Republic, 19–25 August 2001; Simon, P., Ed.; Topology Atlas: Toronto, ON, Canada, April 2002; pp. 331–346.

13. Pestov, V. Topological groups: Where to from here? *Topol. Proc.* **1999**, *24*, 421–502.
14. Pestov, V. *Dynamics of Infinite-Dimensional Groups. The Ramsey-Dvoretzky-Milman Phenomenon;* University Lecture Series, 40; American Mathematical Society: Providence, RI, USA, 2006.
15. Galindo, J. On Group and Semigroup Compactifications of Topological Groups. *Preprint* **2010**.
16. Megrelishvili, M. Fragmentability and representations of flows. *Topol. Proc.* **2003**, *27*, 497–544.
17. Megrelishvili, M. Topological transformation groups: Selected topics. In *Open Problems in Topology II;* Pearl, E., Ed.; Elsevier: Amsterdam, The Netherlands, 2007; pp. 423–438.

axioms

MDPI

Article

Large Sets in Boolean and Non-Boolean Groups and Topology

Ol'ga V. Sipacheva [ORCID]

Department of General Topology and Geometry, Lomonosov Moscow State University, Leninskie Gory 1, Moscow 119991, Russia; o-sipa@yandex.ru

Received: 1 September 2017; Accepted: 23 October 2017; Published: 24 October 2017

Abstract: Various notions of large sets in groups, including the classical notions of thick, syndetic, and piecewise syndetic sets and the new notion of vast sets in groups, are studied with emphasis on the interplay between such sets in Boolean groups. Natural topologies closely related to vast sets are considered; as a byproduct, interesting relations between vast sets and ultrafilters are revealed.

Keywords: large set in a group; vast set; syndetic set; thick set; piecewise syndetic set; Boolean topological group; arrow ultrafilter; Ramsey ultrafilter

---

Various notions of large sets in groups and semigroups naturally arise in dynamics and combinatorial number theory. Most familiar are those of syndetic, thick (or replete), and piecewise syndetic sets. Apparently, the term "syndetic" was introduced by Gottschalk and Hedlund in their 1955 book [1] in the context of topological groups, although syndetic sets of integers have been studied long before (they appear, e.g., in Khintchine's 1934 ergodic theorem). During the past decades, large sets in $\mathbb{Z}$ and in abstract semigroups have been extensively studied. It has turned out that, e.g., piecewise syndetic sets in $\mathbb{N}$ have many attractive properties: they are partition regular (i.e., given any partition of $\mathbb{N}$ into finitely many subsets, at least one of the subsets is piecewise syndetic), contain arbitrarily long arithmetic progressions, and are characterized in terms of ultrafilters on $\mathbb{N}$ (namely, a set is piecewise syndetic it and only if it belongs to an ultrafilter contained in the minimal two-sided ideal of $\beta\mathbb{N}$). Large sets of other kinds are no less interesting, and they have numerous applications to dynamics, Ramsey theory, the ultrafilter semigroup on $\mathbb{N}$, the Bohr compactification, and so on.

Quite recently Reznichenko and the author have found yet another application of large sets. Namely, we introduced special large sets in groups, which we called vast, and applied them to construct a discrete set with precisely one limit point in any countable nondiscrete topological group in which the identity element has nonrapid filter of neighborhoods. Using this technique and special features of Boolean groups, we proved, in particular, the nonexistence of a countable nondiscrete extremally disconnected group in ZFC (see [2]).

In this paper, we study right and left thick, syndetic, piecewise syndetic, and vast sets in groups (although they can be defined for arbitrary semigroups). Our main concern is the interplay between such sets in Boolean groups. We also consider natural topologies closely related to vast sets, which leads to interesting relations between vast sets and ultrafilters.

1. Basic Definitions and Notation

We use the standard notation $\mathbb{Z}$ for the group of integers, $\mathbb{N}$ for the set (or semigroup, depending on the context) of positive integers, and ω for the set of nonnegative integers or the first infinite cardinal; we identify cardinals with the corresponding initial ordinals. Given a set X, by $|X|$ we denote its cardinality, by $[X]^k$ for $k \in \mathbb{N}$, the kth symmetric power of X (i.e., the set of all k-element subsets of X), and by $[X]^{<\omega}$, the set of all finite subsets of X.

Definition 1 (see [3]). *Let G be a group. A set $A \subset G$ is said to be*

(a) *right thick, or simply thick if, for every finite $F \subset S$, there exists a $g \in G$ (or, equivalently, $g \in A$ ([3] Lemma 2.2)) such that $Fg \subset A$;*

(b) *right syndetic, or simply syndetic, if there exists a finite $F \subset G$ such that $G = FA$;*

(c) *right piecewise syndetic, or simply piecewise syndetic, if there exists a finite $F \subset G$ such that FA is thick.*

Left thick, left syndetic, and left piecewise syndetic sets are defined by analogy; in what follows, we consider only right versions and omit the word "right."

Definition 2. *Given a subset A of a group G, we shall refer to the least cardinality of a set $F \subset G$ for which $G = FA$ as the syndeticity index, or simply index (by analogy with subgroups) of A in G. Thus, a set is syndetic if and only if it is of finite index. We also define the thickness index of A as the least cardinality of $F \subset G$ for which FA is thick.*

A set $A \subset \mathbb{Z}$ is syndetic if and only if the gaps between neighboring elements of A are bounded, and $B \subset \mathbb{Z}$ is thick if and only if it contains arbitrarily long intervals of consecutive integers. The intersection of any such sets A and B is piecewise syndetic; clearly, such a set is not necessarily syndetic or thick (although it may as well be both syndetic and thick). The simplest general example of a syndetic set in a group is a coset of a finite-index subgroup.

In what follows, when dealing with general groups, we use multiplicative notation, and when dealing with Abelian ones, we use additive notation.

Given a set A in a group G, by $\langle A \rangle$ we denote the subgroup of G generated by A.

As mentioned, we are particularly interested in Boolean groups, i.e., groups in which all elements are self-inverse. All such groups are Abelian. Moreover, any Boolean group G can be treated as a vector space over the two-element field $\mathbb{Z}_2$; therefore, for some set X (basis), G can be represented as the free Boolean group $B(X)$ on X, i.e., as $[X]^{<\omega}$ with zero $\varnothing$, which we denote by $\mathbf{0}$, and the operation $\triangle$ of symmetric difference: $A \triangle B = (A \cup B) \setminus A \cap B$. We denote this operation treated as the group operation on $B(X)$ by $\blacktriangle$; thus, given $a, b \in B(X) = [X]^{<\omega}$, we have $a \blacktriangle b = a \triangle b$, and given $A, B \subset B(X)$, we have $A \blacktriangle B = \{a \blacktriangle b = a \triangle b : a \in A, b \in B\}$. We identify each $x \in X$ with $\{x\} \in [X]^{<\omega} = B(X)$; thereby, X is embedded in $B(X)$, and the nonzero elements of $B(X)$ are represented by formal sums $x_1 \blacktriangle \cdots \blacktriangle x_n$, where $n \in \mathbb{N}$ and $x_i \in X$, $i \leq n$. Formal sums in which all terms are different are said to be *reduced*. The reduced formal sum representing a given element g of $B(X)$ (that is, a finite subset of X) is determined uniquely up to the order of terms: for $g = \{x_1, \ldots, x_n\}$, this is the sum $x_1 \blacktriangle \cdots \blacktriangle x_n$. We assume that zero is represented by the empty sum. By analogy with the cases of free and free Abelian groups, we refer to the number of terms in the reduced formal sum representing a given element as the *length* of this element. Thus, the length of each element equals its cardinality. Given $n \in \omega$, we use the standard notation $B_n(X)$ for the set of elements of length at most n; thus, $B_0(X) = \{\mathbf{0}\}$, $B_1(X) = X \cup \{\mathbf{0}\}$, and $B(X) = \bigcup_{n \in \omega} B_n(X)$. For the set of elements of length precisely n, where $n \in \mathbb{N}$, we use the notation $B_{=n}(X)$; we have $B_{=n}(X) = B_n(X) \setminus B_{n-1}(X)$. For convenience, pursuing the analogy with free groups, we refer to the terms of the reduced formal sum representing an element g of $B(X)$ as the *letters* of g; thus, each $g \in B(X) = [X]^{<\omega}$ is the set of its letters.

Any free filter $\mathscr{F}$ on an infinite set X determines a topological space $X_{\mathscr{F}} = X \cup \{*\}$ with one nonisolated point $*$; the neighborhoods of this point are $A \cup \{*\}$ for $A \in \mathscr{F}$. The topology of the free Boolean topological group $B(X_{\mathscr{F}}) = [X \cup \{*\}]^{<\omega}$ on this space, that is, the strongest group topology that induces the topology of $X_{\mathscr{F}}$ on $X \cup \{*\}$, is described in detail in [4]. Description II in [4] takes the following form for $X_{\mathscr{F}}$. For each $n \in \mathbb{N}$, we fix an arbitrary sequence Γ of neighborhoods of $*$, that is, $\Gamma = (A_n \cup \{*\})_{n \in \mathbb{N}}$, where $A_n \in \mathscr{F}$, and set

$$U(\Gamma) = \bigcup_{n \in \mathbb{N}} \{x_1 \blacktriangle y_1 \blacktriangle x_2 \blacktriangle y_2 \blacktriangle \cdots \blacktriangle x_n \blacktriangle y_n : x_i, y_i \in A_i \cup \{*\} \text{ for } i \leq n\}.$$

The sets $U(\Gamma)$ form a basis of neighborhoods of zero in $B(X_{\mathscr{F}})$. In particular, the subgroup generated by $(A \cup \{*\}) \vartriangle (A \cup \{*\})$ (and hence the subgroup generated by $A \cup \{*\}$) is a neighborhood of zero for any $A \in \mathscr{F}$. Note that $B(X_{\mathscr{F}})$ contains the abstract free Boolean group $B(X)$ as a subgroup. The topology of $B(X_{\mathscr{F}})$ induces a nondiscrete group topology on $B(X)$; see Section 8 for details.

For the Graev free Boolean topological group (A precise definition can be found in [4]. For aestetic reasons, instead of the standard notation $B_G(\mathscr{F})$ we use $B^G(\mathscr{F})$ in this paper). $B^G(\mathscr{F})$ in which $*$ is identified with zero, a basis of neighborhoods of zero is formed by sets of the form

$$U^G(\Gamma) = \bigcup_{n \in \mathbb{N}} \{x_1 \vartriangle x_2 \vartriangle \cdots \vartriangle x_n : x_i \in A_i \text{ for } i \leq n\}.$$

For spaces of the form $X_{\mathscr{F}}$, the Graev free Boolean topological group is topologically isomorphic to the free Boolean topological group (see [4]).

Clearly, for $n \in \omega$, a set $Y \subset B_{=2n}(X_{\mathscr{F}})$ is a trace on $B_{=2n}(X_{\mathscr{F}})$ of a neighborhood of zero in $B(X_{\mathscr{F}})$ if and only if it contains a set of the form

$$\underbrace{\left((A \cup \{*\}) \vartriangle \cdots \vartriangle (A \cup \{*\})\right)}_{2n \text{ times}} \cap B_{=2n}(X_{\mathscr{F}}) = [A \cup \{*\}]^{2n}.$$

The intersection of a neighborhood of zero in $B(X_{\mathscr{F}})$ with $B_{=k}(X_{\mathscr{F}})$ may be empty for all odd k. Similarly, a set $Y \subset B^G_{=n}(X_{\mathscr{F}})$ is a trace on $B^G_{=n}(X_{\mathscr{F}})$ of a neighborhood of zero in $B^G(X_{\mathscr{F}})$ if and only if it contains a set of the form $\underbrace{\left(A \vartriangle \cdots \vartriangle A\right)}_{n \text{ times}} \cap B^G_{=n}(X_{\mathscr{F}}) = [A]^n$. The intersection of a neighborhood of zero in $B^G(X_{\mathscr{F}})$ with $B^G_n(X_{\mathscr{F}})$ is never empty.

In what follows, we deal with rapid, κ-arrow, and Ramsey filters and ultrafilters.

Definition 3 ([5]). *A filter $\mathscr{F}$ on ω is said to be* rapid *if every function $\omega \to \omega$ is majorized by the increasing enumeration of some element of $\mathscr{F}$, i.e., for any function $a \colon \omega \to \omega$, there exists a strictly increasing function $b \colon \omega \to \omega$ such that $a(i) \leq b(i)$ for all i and $\{b(i) : i \in \omega\} \in \mathscr{F}$.*

Clearly, any filter containing a rapid filter is rapid as well; thus, the existence of rapid filters is equivalent to that of rapid ultrafilters. Rapid ultrafilters are also known as semi-Q-point, or weak Q-point, ultrafilters. Both the existence and nonexistence of rapid ultrafilters is consistent with ZFC (see, e.g., [6,7]).

The notions of κ-arrow and Ramsey filters are closely related to Ramsey theory, more specifically, to the notion of homogeneity with respect to a coloring, or partition. Given a set X and positive integers m and n, by an *m-coloring* of $[X]^n$ we mean any map $c \colon [X]^n \to Y$ of $[X]^n$ to a set Y of cardinality m. Any such coloring determines a partition of $[X]^n$ into m disjoint pieces, each of which is assigned a color $y \in Y$. A set $A \subset X$ is said to be *homogeneous* with respect to c, or *c-homogeneous*, if c is constant on $[A]^n$. The celebrated Ramsey theorem (finite version) asserts that, given any positive integers k, l, and m, there exists a positive integer N such that, for any k-coloring $c \colon [X]^l \to Y$, where $|X| \geq N$ and $|Y| = k$, there exists a c-homogeneous set $A \subset X$ of size m.

We consider κ-arrow and Ramsey filters on any, not necessarily countable, infinite sets. For convenience, we require these filters to be uniform, i.e., nondegenerate in the sense that all of their elements have the same cardinality (equal to that of the underlying set). A filter on a countable set is uniform if and only if it is free.

Definition 4. *Let κ be an infinite cardinal, and let $\mathscr{F}$ be a uniform filter on a set X of cardinality κ.*

(i) *We say that $\mathscr{F}$ is a* Ramsey filter *if, for any 2-coloring $c \colon [X]^2 \to \{0, 1\}$, there exists a c-homogeneous set $A \in \mathscr{U}$.*

(ii) *Given an arbitrary cardinal* $\lambda \leq \kappa$, *we say that* $\mathscr{F}$ *is a* λ-*arrow filter if, for any 2-coloring* $c\colon [X]^2 \to \{0,1\}$, *there exists either a set* $A \in \mathscr{F}$ *such that* $c([A]^2) = \{0\}$ *or a set* $S \subset X$ *with* $|S| \geq \lambda$ *such that* $c([S]^2) = \{1\}$.

Any filter $\mathscr{F}$ on X which is Ramsey or λ-arrow for $\lambda \geq 3$ is an ultrafilter. Indeed, let $S \subset X$ and consider the coloring $c\colon [X]^2 \to \{0,1\}$ defined by

$$c(\{x,y\}) = \begin{cases} 0 & \text{if } x,y \in S \text{ or } x,y \in X \setminus S, \\ 1 & \text{otherwise.} \end{cases}$$

Clearly, any c-homogeneous set containing more than two points is contained entirely in S or in $X \setminus S$; therefore, either S or $X \setminus S$ belongs to $\mathscr{F}$, so that $\mathscr{F}$ is an ultrafilter.

According to Theorem 9.6 in [8], if $\mathscr{U}$ is a Ramsey ultrafilter on X, then, for any $n < \omega$ and any 2-coloring $c\colon [X]^n \to \{0,1\}$, there exists a c-homogeneous set $A \in \mathscr{U}$.

It is easy to see that if $\mathscr{F}$ is λ-arrow, then, for any $A \in \mathscr{F}$ and any $c\colon [A]^2 \to \{0,1\}$, there exists either a set $B \in \mathscr{F}$ such that $B \subset A$ and $c([B]^2) = \{0\}$ or a set $S \subset A$ with $|S| \geq \lambda$ such that $c([S]^2) = \{1\}$.

In [9], where k-arrow ultrafilters for finite k were introduced, it was shown that the existence of a 3-arrow (ultra)filter on ω implies that of a P-point ultrafilter; therefore, the nonexistence of κ-arrow ultrafilters for any $\kappa \geq 3$ is consistent with ZFC (see [10]).

On the other hand, the continuum hypothesis implies the existence of k-arrow ultrafilters on ω for any $k \leq \omega$. To formulate a more delicate assumption under which k-arrow ultrafilters exist, we need more definitions. Given a free (=uniform) filter $\mathscr{F}$ on ω, a set $B \subset \omega$ is called a *pseudointersection* of $\mathscr{F}$ if the complement $A \setminus B$ is finite for all $A \in \mathscr{F}$. The *pseudointersection number* $\mathfrak{p}$ is the smallest size of a free filter on ω which has no infinite pseudointersection. It is easy to show that $\omega_1 \leq \mathfrak{p} \leq 2^\omega$, so that, under the continuum hypothesis, $\mathfrak{p} = 2^\omega$. It is also consistent with ZFC that, for any regular cardinals κ and λ such that $\omega_1 \leq \kappa \leq \lambda$, $2^\omega = \lambda$ and $\mathfrak{p} = \kappa$ (see [11] Theorem 5.1). It was proved in [9] that, under the assumption $\mathfrak{p} = 2^\omega$ (which is referred to as P(c) in [9]), there exist κ-arrow ultrafilters on ω for all $\kappa \leq \omega$. Moreover, for each $k \in \mathbb{N}$, there exists a k-arrow ultrafilter on ω which is not $(k+1)$-arrow, and there exists an ultrafilter which is k-arrow for each $k \in \mathbb{N}$ but is not Ramsey and hence not ω-arrow ([9] Theorems 2.1 and 4.10).

In addition to the free group topology of Boolean groups on spaces generated by filters, we consider the *Bohr topology* on arbitrary abstract and topological groups. This is the weakest group topology with respect to which all homomorphisms to compact topological groups are continuous, or the strongest totally bounded group topology; the Bohr topology on an abstract group (without topology) is defined as the Bohr topology on this group endowed with the discrete topology.

Finally, we need the definition of a minimal dynamical system.

Definition 5. *Let* G *be a monoid with identity element* e. *A pair* $(X, (T_g)_{g \in G})$, *where* X *is a topological space and* $(T_g)_{g \in G}$ *is a family of continuous maps* $X \to X$ *such that* T_e *is the identity map and* $T_{gh} = T_g \circ T_h$ *for any* $g, h \in G$, *is called a topological dynamical system. Such a system is said to be minimal if no proper closed subset of* X *is* T_g-*invariant for all* $g \in G$.

We sometimes identify sequences with their ranges.

All groups considered in this paper are assumed to be infinite, and all filters are assumed to have empty intersection, i.e., to contain the Fréchet filter of all cofinite subsets (and hence be free).

2. Properties of Large Sets

We begin with well-known general properties of large sets defined above. Let G be a group.

Property 1. *A set $A \subset G$ is thick if and only if the family $\{gA : g \in G\}$ of all translates of A has the finite intersection property.*

Indeed, this property means that, for every finite subset F of G, there exists an $h \in \bigcap_{g \in F} g^{-1} A$, and this, in turn, means that $gh \in A$ for each $g \in F$, i.e., $Fh \subset A$.

Property 2. *([3] Theorem 2.4) The family of syndetic sets A set A is syndetic if and only if A intersects every thick set, or, equivalently, if its complement $G \setminus A$ is not thick.*

Given a family $\mathcal{F}$ of subsets of a set X, the *dual* family $\mathcal{F}^*$ is defined as $\mathcal{F}^* = \{A \subset X : A \cup B \neq \varnothing$ for any $B \in \mathcal{F}\}$ (see, e.g., [12]). Thus, Property 2 says that the family of syndetic sets is dual to that of thick sets. The next property is an obvious reformulation of this fact.

Property 3. *A set A is thick if and only if A intersects every syndetic set, or, equivalently, if its complement $G \setminus A$ is not syndetic. In other words, the family of thick sets is dual to that of syndetic sets.*

Property 4. *([3] Theorem 2.4) A set A is piecewise syndetic if and only if there exists a syndetic set B and a thick set C such that $A = B \cap C$.*

Property 5. *([13] Theorem 4.48) A set A is thick if and only if*

$$\overline{A}^{\beta G} = \{p \in \beta G : A \in p\}$$

(the closure of A in the Stone–Čech compactification βG of G with the discrete topology) contains a left ideal of the semigroup βG.

Property 6. *([13] Theorem 4.48) A set A is syndetic if and only if every left ideal of βG intersects $\overline{A}^{\beta G}$.*

Property 7. *The families of thick, syndetic, and piecewise syndetic sets are closed with respect to taking supersets.*

Property 8. *Thickness, syndeticity, and piecewise syndeticity are invariant under both left and right translations.*

Property 9. *([3] Theorem 2.5) Piecewise syndeticity is* partition regular, *i.e., whenever a piecewise syndetic set is partitioned into finitely many subsets, one of these subsets is piecewise syndetic.*

Property 10. *([3] Theorem 2.4) For any thick set $A \subset G$, there exists an infinite sequence $B = (b_n)_{n \in \mathbb{N}}$ in G such that*

$$\mathrm{FP}(B) = \{x_{n_1} x_{n_2} \ldots x_{n_k} : k, n_1, n_2, \ldots, n_k \in \mathbb{N},\ n_1 < n_2 < \cdots < n_k\}$$

is contained in A.

Property 11. *Any IP^*-set in G, i.e., a set intersecting any infinite set of the form $\mathrm{FP}(B)$, is syndetic. This immediately follows from Properties 2 and 10.*

3. Vast Sets

As mentioned at the beginning of this section, in [2], Reznichenko and the author introduced a new (Later, we have found out that similar subsets of $\mathbb{Z}$ had already been used in [14]: the Δ_n^*-sets considered there and n-vast subsets of $\mathbb{Z}$ are very much alike). class of large sets, which we called vast; they have played the key role in our construction of nonclosed discrete subsets in topological groups.

Definition 6. *We say that a subset A of a group G is* vast *in G if there exists a positive integer m such that any m-element set F in G contains a two-element subset D for which $D^{-1}D \subset A$. The least number m with this property is called the* vastness *of A.*

    We shall refer to vast sets of vastness m as m-vast sets.

In a similar manner, κ-vast sets for any cardinal κ can defined.

Definition 7. *Given a cardinal κ, we say that a subset A of a group G is* κ-vast *in G if any set $S \subset G$ with $|S| = \kappa$ contains a two-element subset D for which $D^{-1}D \subset A$.*

The notions of an ω-vast and a k-vast set are very similar to but different from those of Δ^*- and Δ_k^*-sets. Δ^*-Sets were introduced and studied in [3] for arbitrary semigroups, and Δ_k^*-sets with $k \in \mathbb{N}$ were defined in [14] for the case of $\mathbb{Z}$.

Definition 8. *Given a finite or countable cardinal κ and a sequence $(g_n)_{n \in \kappa}$ in a group G, we set*

$$\Delta_I\big((g_n)_{n \in \kappa}\big) = \{x \in G : \text{there exist } m < n < \kappa \text{ such that } x = g_m^{-1}g_n\}$$

and

$$\Delta_D\big((g_n)_{n \in \kappa}\big) = \{x \in G : \text{there exist } m < n < \kappa \text{ such that } x = g_n g_m^{-1}\}.$$

    A subset of a group G is called a right (left) Δ_κ^*-set *if it intersects $\Delta_I\big((g_n)_{n \in \kappa}\big)$ (respectively, $\Delta_D\big((g_n)_{n \in \kappa}\big)$) for any one-to-one sequence $(g_n)_{n \in \kappa}$ in G; Δ_ω^*-sets are referred to as Δ^*-sets.*

Remark. *For any one-to-one sequence $S = (g_n)_{n \in \kappa}$ in a Boolean group with zero **0**, we have $\Delta_I(S) = \Delta_D(S) = (S \triangle S) \setminus \{\mathbf{0}\}$. Hence any κ-vast set in such a group is a right and left Δ_κ^*-set. Moreover, the only difference between Δ_κ^*- and κ-vast sets in a Boolean group is in that the latter must contain **0**, i.e., a set A in such a group is vast if and only if $\mathbf{0} \in A$ and A is a Δ_κ^*-set.*

The most obvious feature distinguishing vastness among other notions of largeness is symmetry (vastness has no natural right and left versions). In return, translation invariance is sacrificed. Thus, in studying vast sets, it makes sense to consider also their translates.

    Clearly, a 2-vast set in a group must coincide with this group. The simplest nontrivial example of a vast set is a subgroup of finite index n; its vastness equals $n + 1$ (any $(n + 1)$-element subset has two elements x and y in the same coset, and both $x^{-1}y$ and $y^{-1}x$ belong to the subgroup).

    It seems natural to refine the definition of vast sets by requiring $A \cap F^{-1}F$ to be of prescribed size rather than merely nontrivial. However, this (and even a formally stronger) requirement does not introduce anything new.

Proposition 1 ([2], Proposition 1.1). *For any vast set A in a group G and any positive integer n, there exists a positive integer m such that any m-element set F in G contains an n-element subset E for which $E^{-1}E \subset A$.*

Indeed, considering the coloring $c \colon [G]^2 \to \{0, 1\}$ defined by $c(\{x, y\}) = 1 \iff x^{-1}y, y^{-1}x \in A$ and applying the finite Ramsey theorem, we find a c-homogeneous set E of size n (provided that m is large enough). If n is no smaller than the vastness of A (which we can assume without loss of generality), then $c([E]^2) = \{1\}$.

    There is yet another important distinguishing feature of vast sets, namely, the finite intersection property. Neither thick, syndetic, nor piecewise syndetic sets have this property (Indeed, the disjoint sets of even and odd numbers are syndetic in $\mathbb{Z}$, and $\bigcup_{i \geq 0}[2^{2i}, 2^{2i+1}) \cap \mathbb{Z}$ and $\bigcup_{i \geq 1}[2^{2i-1}, 2^i)$ are thick). The following theorem is valid.

Theorem 1 ([2]). *Let G be a group .*

(i)   If $A \subset G$ is vast, then so is A^{-1}.
(ii) If $A \subset B \subset G$ and A is vast, then so is B.
(iii) If $A \subset G$ and $B \subset G$ are vast, then so is $A \cap B$.

Assertions (i) and (ii) are obvious, and (iii) follows from Proposition 1.

Proposition 2. *If G is a group, $S \subset G$, and $S \cap (SS \cup S^{-1}S^{-1}) = \varnothing$, then $G \setminus S$ is 3-vast.*

Proof. Take any three different elements $a, b, c \in G$. We must show that the identity element e belongs to $G \setminus S$ (which is true by assumption) and either $(a^{-1}b)^{\pm 1} \in G \setminus S$, $(b^{-1}c)^{\pm 1} \in G \setminus S$, or $(c^{-1}a)^{\pm 1} \in G \setminus S$. Assume that, on the contrary, $(a^{-1}b)^\varepsilon \in S$ (i.e., $a^{-1}b \in S^\varepsilon$), $b^{-1}c \in S^\delta$, and $c^{-1}a \in S^\gamma$ for some $\varepsilon, \delta, \gamma \in \{-1, 1\}$. At least two of the three numbers ε, δ, and γ are equal. Suppose for definiteness that $\varepsilon = \delta$. Then we have $c^{-1}a = c^{-1}bb^{-1}a \in S^{-\varepsilon}S^{-\varepsilon}$, which contradicts the assumption $S \cap (S^2 \cup S^{-2}) = \varnothing$. □

We see that the family of vast sets in a group resembles, in some respects, a base of neighborhoods of the identity element for a group topology. However, as we shall see in the next section, it does not generate a group topology even in a Boolean group: any Boolean group has a 3-vast subset A containing no set of the form $B \bigtriangleup B$ for vast B. On the other hand, many groups admit of group topologies in which all neighborhoods of the identity element are vast; for example, such are topologies generated by normal subgroups of finite index. A more precise statement is given in the next section. Before turning to related questions, we consider how vast sets fit into the company of other large sets.

We begin with a comparison of vast and syndetic sets.

Proposition 3 (see [2] Proposition 1.7). *Let G be any group with identity element e. Any vast set A in G is syndetic, and its syndeticity index is less than its vastness.*

Proof. Let n denote the vastness of A. Take a finite set $F \subset G$ with $|F| = n - 1$ such that $x^{-1}y \notin A$ or $y^{-1}x \notin A$ for any different $x, y \in F$. Pick any $g \in G \setminus F$. Since $|F \cup \{g\}| = n$, it follows that $x^{-1}g \in A$ and $g^{-1}x \in A$ for some $x \in F$, whence $g \in xA$, i.e., $G \setminus F \subset FA$. By definition, the identity element of G belongs to A, and we finally obtain $G = FA$. □

Examples of nonvast syndetic sets are easy to construct: any coset of a finite-index subgroup in a group is syndetic, while only one of them (the subgroup itself) is vast. However, the existence of syndetic sets with nonvast translates is not so obvious. An example of such a set in $\mathbb{Z}$ can be extracted from [14].

Example 1. *There exists a syndetic set in $\mathbb{Z}$ such that none of its translates is vast. This is, e.g., the set constructed in ([14] Theorem 4.3). Namely, let $C = \{0, 1\}^{\mathbb{Z}}$, and let $\tau : C \to C$ be the shift, i.e., the map defined by $\tau(f)(n) = f(n + 1)$ for $f \in C$. It was proved in ([14] Theorem 4.3) that if $M \subset C$ is a minimal closed τ-invariant subset (Then the support of each $f \in M$ is syndetic in $\mathbb{Z}$ (see, e.g., [15])). and the dynamical system $(M, (\tau^n)_{n \in \mathbb{Z}})$ satisfies a certain condition (Namely, is weakly mixing; see, e.g., [15]) , then the support of any $f \in M$ is syndetic but not piecewise Bohr; the latter means that it cannot be represented as the intersection of a thick set and a set having nonempty interior in the Bohr topology on $\mathbb{Z}$. Clearly, any translate of supp f has these properties as well. On the other hand, according to Theorem II in [14], any Δ_n^*-set in $\mathbb{Z}$ (i.e., any set intersecting the set of differences $\{k_j - k_i : i < j \leq n\}$ for each n-tuple $(k_1, \ldots, k_n)$ of different integers) is piecewise Bohr. Since every n-vast set is a Δ_n^*-set, it follows that the translates of supp f cannot be vast.*

Bearing in mind our particular interest in Boolean groups, we also give a similar example for a Boolean group.

Example 2. *We construct a syndetic set in the Boolean group $B(\mathbb{Z})$ with nonvast translates. Let S be a syndetic set in $\mathbb{Z}$ all of whose translates are not Δ_n^*-sets for all n (see Example 1). By definition, $\mathbb{Z} = \bigcup_{k \leq r}(s_k + S)$ for some $r \in \mathbb{N}$ and different $s_1, \ldots, s_r \in \mathbb{Z}$. We set*

$$S_k' = \{x_1 \triangle \cdots \triangle x_n : n \in \mathbb{N}, x_i \in \mathbb{Z} \text{ for } i \leq n, x_i \neq x_j \text{ for } i \neq j,$$

$$\{x_1, \ldots, x_n\} \cap \{s_1, \ldots, s_r\} = \{s_k\}, \sum_{i \leq n} x_i \in 2s_k + S\}, \qquad k \leq r,$$

and

$$S' = \bigcup_{k \leq r} S_k'.$$

We have

$$s_k \triangle S_k' = \{x_1 \triangle \cdots \triangle x_n : n \in \mathbb{N}, x_i \in \mathbb{Z} \text{ for } i \leq n, x_i \neq x_j \text{ for } i \neq j,$$

$$\{x_1, \ldots, x_n\} \cap \{s_1, \ldots, s_r\} = \varnothing, \sum_{i \leq n} x_i \in s_k + S\}, \qquad k \leq r.$$

Since $\bigcup_{k \leq r}(s_k + S) = \mathbb{Z}$, it follows that

$$\bigcup_{k \leq r}(s_k \triangle S') \subset \{x_1 \triangle \cdots \triangle x_n : n \in \mathbb{N}, x_i \in \mathbb{Z} \text{ for } i \leq n, \{x_1, \ldots, x_n\} \cap \{s_1, \ldots, s_r\} = \varnothing\}.$$

Obviously, the set on the right-hand side of this inclusion is syndetic; therefore, so is S'.

Let us show that no translate of S' is vast. Suppose that, on the contrary, $k, n \in \mathbb{N}$, $z_1, \ldots, z_k \in \mathbb{Z}$, $w = z_1 \triangle \cdots \triangle z_k$, and $w \triangle S'$ is n-vast. Take any different $k_1, \ldots, k_n \in \mathbb{Z}$ larger than the absolute values of all elements of w (which is a finite subset of $\mathbb{Z}$) and of all s_i, $i \leq r$. We set

$$F = \{k_1, k_1 \triangle (-k_1) \triangle k_2, k_1 \triangle (-k_1) \triangle k_2 \triangle (-k_2) \triangle k_3,$$

$$\ldots, k_1 \triangle (-k_1) \triangle k_2 \triangle (-k_2) \triangle \cdots \triangle k_{n-1} \triangle (-k_{n-1}) \triangle k_n\}.$$

Suppose that there exist different $x, y \in F$ for which $x \triangle y \in w \triangle S'$, i.e., there exist $i, j \leq n$ for which $i < j$ and

$$k_1 \triangle (-k_1) \triangle \cdots \triangle k_{i-1} \triangle (-k_{i-1}) \triangle k_i \triangle k_1 \triangle (-k_1) \triangle \cdots \triangle k_{j-1} \triangle (-k_{j-1}) \triangle k_j$$

$$= k_i \triangle k_i \triangle (-k_i) \triangle k_{i+1} \triangle (-k_{i+1}) \triangle \cdots \triangle k_{j-1} \triangle (-k_{j-1}) \triangle k_j = w \triangle s \in w \triangle S',$$

where s is an element of S' and hence belongs to S_l' for some $l \leq r$, which means, in particular, that s contains precisely one of the letters $s_1, \ldots, s_r$, namely, s_l. There are no such letters among $\pm k_i, \ldots, \pm k_{j-1}, k_j$. Therefore, one of the letters z_m (say z_1) is s_l. The other letters of w do not equal $\pm k_i, \ldots, \pm k_{j-1}, k_j$ either and, therefore, are canceled with letters of $s \in S'$ in $w + s$. By the definition of the set S' containing s, one letter of w (namely, $z_1 = s_l$) belongs to the set $\{s_1, \ldots, s_r\}$ and the other letters do not. Since the sum (in $\mathbb{Z}$) of the integer-letters of s belongs to $2s_l + S$ (by the definition of S_l') and $s_l = z_1$, it follows that the sum of letters of $w + s$ belongs to $S + z_1 - z_2 - \cdots - z_k$ and the letter z_1 is determined uniquely for the given element w. To obtain a contradiction, it remains to recall that the translates of S (in particular, $S + z_1 - z_2 - \cdots - z_k$) are not Δ_n^*-sets in $\mathbb{Z}$ and choose $k_1, \ldots, k_n$ so that $\{k_j - k_i : i < j \leq n\} \cap (S + z_1 - z_2 - \cdots - z_k) = \varnothing$.

Example 3. *There exist vast sets which are not thick and thick sets which are not vast. Indeed, as mentioned, any proper finite-index group is vast, but it cannot be thick by the first property in the list of properties of large sets given above.*

An example of a nonvast thick set is, e.g., any thick nonsyndetic set. In an infinite Boolean group G, such a set can be constructed as follows. Take any basis X in G (so that $G = B(X)$), fix any nonsyndetic thick set T in $\mathbb{N}$ (say $T = \bigcup_n([a_n, b_n] \cap \mathbb{N}$, where the a_n and b_n are numbers such that the $b_n - a_n$ and the $a_{n+1} - b_n$ increase without bound), and consider the set

$$A = \{x_1 \triangle \cdots \triangle x_n \in B(X) : n \in T, x_i \in X \text{ for } i \leq n, x_i \neq x_j \text{ for } i \neq j\}$$

of all elements in $B(X)$ whose lengths belong to T. The thickness of this set is obvious (by the same Property 1), because the translate of A by any element $g \in B(X)$ of any length l surely contains all elements whose lengths belong to $\bigcup_n([a_n + l, b_n - l] \cap \mathbb{N}) \subset T$ and, therefore, intersects A. However, A is not vast, because it misses all elements whose lengths belong to the set $\bigcup_n((b_n, a_{n+l}) \cap \mathbb{N})$. The last set contains at least one even positive integer $2k$. It remains to choose different points $x_1, x_2, \ldots$ in X, set $B = \{x_{kn+1} \triangle x_{kn+2} \cdots \triangle x_{kn+k} : n \in \omega\}$, and note that all nonempty elements of $B \triangle B$ have length $2k$. Therefore, A is disjoint from $B \triangle B$ (much more from $F \triangle F$ for any finite $F \subset B$). Note that the translates of A are not vast either, because both thickness and (non)syndeticity are translation invariant.

Proposition 4. *Let G be any group with identity element e.*

  (i)   *If a set A in G is 3-vast, then $(G \setminus A)^{-1}(G \setminus A) \subset A$.*
 (ii) *If a set A in G is 3-vast, then either $AA^{-1} = G$ or A is a subgroup of index 2.*

Proof. (i) Suppose that A is a 3-vast subset of a group G with identity element e. Take any different $x, y \notin A$ (if there exist no such elements, then there is nothing to prove). By definition, the set $\{x, y, e\}$ contains a two-element subset D for which $D^{-1}D \subset A$. Clearly, $D \neq \{x, e\}$ and $D \neq \{y, e\}$. Therefore, $x^{-1}y \in A$ and $y^{-1}x \in A$ (and $e \in A$, too), whence $(G \setminus A)^{-1}(G \setminus A) \subset A$.

    (ii) If $AA^{-1} \neq G$, then there exists a $g \in G$ for which $gA \cap A = \varnothing$. If A is, in addition, 3-vast, then (ii) implies $(gA)^{-1}gA = A^{-1}A \subset A$, which means that A is a subgroup of G. According to (i), A is syndetic of index at most 2; in fact, its index is precisely 2, because A does not coincide with G. $\square$

4. Quotient Sets

In [3] sets of the form AA^{-1} or $A^{-1}A$ were naturally called *quotient sets*. We shall refer to the former as *right quotient sets* and to the latter as *left quotient sets*. Thus, a set in a group G is m-vast if it intersects nontrivially the left quotient set of any m-element subset of G. Quotient sets play a very important role in combinatorics, and their interplay with large sets is quite amazing.

    First, the passage to right quotient sets annihilates the difference between syndetic and piecewise syndetic sets.

Theorem 2 (see [3], Theorem 3.9). *For each piecewise syndetic subset A of a group G, there exists a syndetic subset B of G such that $BB^{-1} \subset AA^{-1}$ and the syndeticity index of B does not exceed the thickness index of A.*

    Briefly, the construction of B given in [3] is as follows: we take a finite set T such that TA is thick and, for each finite $F \subset G$, let $\Phi_F = \{\varphi \in T^G : \bigcap_{x \in F} x^{-1}\varphi(x)A \neq \varnothing\}$. Then we pick φ^* in the intersection of all Φ_F (which exists since the product space T^G is compact) and let $B = \{\varphi^*(x)^{-1}x : x \in G\}$. Since $\varphi^*(G) \subset T$, it follows that $TB = G$, which means that B is syndetic and its index does not exceed $|T| = t$. Moreover, for any finite $F \subset B$, there exists a $g \in G$ such that $Fg \subset A$, and this implies $BB^{-1} \subset AA^{-1}$.

    In Theorem 2, right quotient sets cannot be replaced by left ones: there are examples of piecewise syndetic sets A such that $A^{-1}A$ does not contain $B^{-1}B$ for any syndetic B. One of such examples is provided by the following theorem.

Theorem 3. *The following assertions hold.*

(i) If a subset A of a group G is syndetic of index s, then $A^{-1}A$ is vast, and its vastness does not exceed $s+1$.

(ii) If a subset A of an Abelian group G is piecewise syndetic of thickness index t, then $A - A$ is vast, and its vastness does not exceed $t+1$.

(iii) There exists a group G and a thick (in particular, piecewise syndetic) set $A \subset G$ such that $A^{-1}A$ is not vast and, therefore, does not contain $B^{-1}B$ for any syndetic set.

(iv) If a subset A of a group G is thick, then $AA^{-1} = G$.

Proof. (i) Suppose that $FA = G$, where $F = \{g_1, \dots, g_s\}$. Any $(s+1)$-element subset of G has at least two points x and y in the same "coset" $g_i A$. We have $x = g_i a'$ and $y = g_i a''$, where $a', a'' \in A$. Thus, $x^{-1}y, y^{-1}x \in A^{-1}A$.

Assertion (ii) follows immediately from (i) and Theorem 2.

Let us prove (iii). Consider the free group G on two generators a and b and let A be the set of all words in G whose last letter is a. Then A is thick (given any finite $F \subset G$, we have $Fa^n \subset A$ for sufficiently large n). Clearly, all nonidentity words in $A^{-1}A$ contain a or a^{-1}. Therefore, if $F \subset G$ consists of words of the form b^n, then the intersection $F^{-1}F \cap A^{-1}A$ is trivial, so that $A^{-1}A$ is not vast.

Finally, to prove (iv), take any $g \in G$. We have $A \cap gA \neq \varnothing$ (by Property 1 in our list of properties of large sets). This means that $g \in AA^{-1}$. □

We see that the right quotient sets AA^{-1} of thick sets A are utmostly large, while the left quotient sets $A^{-1}A$ may be rather small. In the Abelian case, the difference sets of all thick sets coincide with the whole group.

It is natural to ask whether condition (i) in Theorem 3 characterizes vast sets in groups. In other words, given any vast set A in a group, does there exist a syndetic (or, equivalently, piecewise syndetic) set B such that $B^{-1}B \subset A$ (or $BB^{-1} \subset A$)? The answer is no, even for thick 3-vast sets in Boolean groups. The idea of the following example was suggested by arguments in paper [14] and in John Griesmer's note [16], where the group $\mathbb{Z}$ was considered.

Example 4. *Let G be a countable Boolean group with zero $\mathbf{0}$. Any such group can be treated as the free Boolean group on $\mathbb{Z}$. We set*

$$A = G \setminus \{m \triangle n = \{m, n\} : m, n \in \mathbb{Z}, m < n, n - m = k^3 \text{ for some } k \in \mathbb{N}\}.$$

Clearly, A is thick (if $F \subset G$ is finite and an element $g \in G$ is sufficiently long, then all elements in the set $F \triangle g$ have more than two letters and, therefore, belong to A). Let us prove that A is 3-vast. Take any different $a, b, c \in G$. We must show that $a \triangle b \in A$, $b \triangle c \in A$, or $a \triangle c \in A$. We can assume that $c = \mathbf{0}$; otherwise, we translate a, b, and c by c, which does not affect the Boolean sums. Thus, it suffices to show that, given any different nonzero $x, y \notin A$, we have $x \triangle y \in A$. The condition $x, y \notin A$ means that $x = \{k, l\}$, where $k < l$ and $l - k = r^3$ for some $r \in \mathbb{Z}$, and $y = \{m, n\}$, where $m < n$ and $n - m = s^3$ for some $s \in \mathbb{Z}$. Suppose for definiteness that $n > l$ or $n = l$ and $m > k$. If $x \triangle y \notin A$, then either $k = m$ and $l - n = t^3$ for some $t \in \mathbb{N}$, $l = m$ and $n - k = t^3$ for some $t \in \mathbb{N}$, or $l = n$ and $m - k = t^3$ for some $t \in \mathbb{N}$. In the first case, we have $l - k = l - n + n - m$, i.e., $r^3 = t^3 + s^3$; in the second, we have $n - k = n - m + l - k$, i.e., $t^3 = s^3 + r^3$; and in the third, we have $l - k = n - m + m - k$, i.e., $r^3 = s^3 + t^3$. In any case, we obtain a contradiction with Fermat's theorem.

It remains to prove that there exists no syndetic (and hence no piecewise syndetic) $B \subset G$ for which $B \triangle B \subset A$. Consider any syndetic set B. Let $F = \{f_1, \dots, f_k\} \subset G$ be a finite set for which $FB = G$, and let m be the maximum absolute value of all letters of elements of F (recall that all letters are integers). To each $n \in \mathbb{Z}$ with $|n| > m$ we assign an element $f_i \in F$ for which $n \in f_i \triangle B$; if there are several such elements, then we choose any of them. Thereby, we divide the set of all integers with absolute value larger than m into k pieces $I_1, \dots, I_k$. To accomplish our goal, it suffices to show that there is a piece I_i containing two integers r and s such that $r - s = z^3$ for some $z \in \mathbb{Z}$. Indeed, in this case, we have $r \in f_i \triangle B$ and $s \in f_i \triangle B$, so that $r \triangle s \in B \triangle B$. On the other hand, $r \triangle s \notin A$.

From now on, we treat the pieces $I_1, \ldots, I_k$ as subsets of $\mathbb{Z}$. We have $\mathbb{Z} = \{-m, -m+1, \ldots, 0, 1, \ldots, m\} \cup I_1 \cup \cdots \cup I_k$. Since piecewise syndeticity is partition regular (see Property 9 of large sets), one of the sets I_i, say I_l, is piecewise syndetic. Therefore, by Theorem 2 , $I_l - I_l \supset S - S$ for some syndetic set $S \subset \mathbb{Z}$.

Let $d^*(S)$ denote the upper Banach density of S, i.e.,

$$d^*(S) = \limsup_{|I| \to \infty} \frac{|S \cap I|}{|I|},$$

where I ranges over all intervals of $\mathbb{Z}$. The syndeticity of S in $\mathbb{Z}$ implies the existence of an $N \in \mathbb{N}$ such that every interval of integers longer than N intersects S. Clearly, we have $d^*(S) \geq 1/N$. Proposition 3.19 in [15] asserts that if X is a set in $\mathbb{Z}$ of positive upper Banach density and $p(t)$ is a polynomial taking on integer values at the integers and including 0 in its range on the integers, then there exist $x, y \in X$, $x \neq y$, and $z \in \mathbb{Z}$ such that $x - y = p(z)$ (as mentioned in [15], this was proved independently by Sárközy). Thus, there exist different $x, y \in S$ and a $z \in \mathbb{Z}$ for which $x - y = z^3$. Since $S - S \subset I_l - I_l$, it follows that $z^3 = r - s$ for some $r, s \in I_l$, as desired.

5. Large Sets and Topology

In the context of topological groups, quotient sets arise again, because for each neighborhood U of the identity element, there must exist a neighborhood V such that $V^{-1}V \subset U$ and $VV^{-1} \subset U$. Thus, if we know that a group topology consists of piecewise syndetic sets, then, in view of Theorem 2, we can assert that all open sets are syndetic, and so on. Example 4 shows that if G is any countable Boolean topological group and all 3-vast sets are open in G, then some nonempty open sets in this group are not piecewise syndetic. Thus, all syndetic or piecewise syndetic subsets of a group G do not generally form a group topology. Even their quotient (difference in the Abelian case) sets are insufficient; however, it is known that double difference sets of syndetic (and hence piecewise syndetic) sets in Abelian groups are neighborhoods of zero in the Bohr topology (It follows, in particular, that, given any piecewise syndetic set A in an Abelian group, there exists an infinite sequence of vast sets $A_1, A_2, \ldots$ such that $A_1 - A_1 \subset A + A - A - A$ and $A_{n+1} - A_{n+1} \subset A_n$ for all n (because all Bohr open sets are syndetic)). These and many other interesting results concerning a relationship between Bohr open and large subsets of abstract and topological groups can be found in [17,18]. As to group topologies in which all open sets are large, the situation is very simple.

Theorem 4. *For any topological group G with identity element e, the following conditions are equivalent:*

  (i)   *all neighborhoods of e in G are piecewise syndetic;*
 (ii)   *all open sets in G are piecewise syndetic;*
(iii)   *all neighborhoods of e in G are syndetic;*
 (iv)   *all open sets in G are syndetic;*
  (v)   *all neighborhoods of e in G are vast;*
 (vi)   *G is totally bounded.*

Proof. The equivalences (i) $\Leftrightarrow$ (ii) and (iii) $\Leftrightarrow$ (iv) follow from the obvious translation invariance of piecewise syndeticity and syndeticity. Theorem 2 implies (i) $\Leftrightarrow$ (iii), Theorem 3 (i) implies (iii) $\Rightarrow$ (v), and Proposition 3 implies (v) $\Rightarrow$ (iii). The implication (iii) $\Rightarrow$ (i) is trivial. Finally, (vi) $\Leftrightarrow$ (iii) by the definition of total boundedness. □

Thus, the Bohr topology on a (discrete) group is the strongest group topology in which all open sets are syndetic (or, equivalently, piecewise syndetic, or vast).

For completeness, we also mention the following corollary of Theorem 3 and Theorem 3.12 in [3], which relates vast sets to topological dynamics.

Corollary 1. *If G is an Abelian group with zero 0, X is a compact Hausdorff space, and* $(X, (T_g)_{g \in G})$ *is a minimal dynamical system, then the set* $\{g \in G : U \cap T_g^{-1}U \neq \varnothing\}$ *is vast for every nonempty open subset U of X.*

6. Vast and Discrete Sets in Topological Groups

As mentioned above, vast sets were introduced in [2] to construct discrete sets in topological groups. Namely, given a countable topological group G whose identity element e has nonrapid filter $\mathscr{F}$ of neighborhoods, we can construct a discrete set with precisely one limit point in this group as follows. The nonrapidness of $\mathscr{F}$ means that, given any sequence $(m_n)_{n \in \mathbb{N}}$ of positive integers, there exist finite sets $F_n \subset G$, $n \in \mathbb{N}$, such that each neighborhood of e intersects some F_n in at least m_n points (see [7] Theorem 3 (3)). Thus, if we have a decreasing sequence of closed m_n-vast sets A_n in G such that $\cap A_n = \{e\}$, then the set

$$D = \bigcup_{n \in \mathbb{N}} \{a^{-1}b : a \neq b, a, b \in F_n, a^{-1}b \in A_n\}$$

is discrete (because $e \notin D$ and each $g \in G \setminus \{e\}$ has a neighborhood of the form $G \setminus A_n$ which contains only finitely many elements of D), and e is the only limit point of D (because, given any neighborhood U of e, we can take a neighborhood V such that $V^{-1}V \subset U$; we have $|V \cap F_n| \geq m_n$ for some n, and hence $(V \cap F_n)^{-1}(V \cap F_n) \cap A_n \neq \varnothing$, so that $U \cap D \neq \varnothing$). It remains to, first, find a family of closed vast sets with trivial intersection and, secondly, make it decreasing.

The former task is easy to accomplish in any topological group: by Proposition 2, in any topological group G, the complements to open neighborhoods gU of all $g \in G$ satisfying the condition $gU \cap (U^2 \cup (U^{-1})^2) = \varnothing$ form a family of closed 3-vast sets with trivial intersection. In countable topological groups, the latter task can be easily accomplished as well: the above family can be made decreasing by using Theorem 1, according to which the family of vast sets has the finite intersection property. Unfortunately, no similar argument applies in the uncountable case, because countable intersections of vast sets may be very small. Thus, in $\mathbb{Z}_2^\omega$, the intersection of the 3-vast sets $H_n = \{f \in \mathbb{Z}_2^\omega : f(n) = 0\}$ (each of which is a subgroup of index 2 open in the product topology) is trivial.

7. Large Sets in Boolean Groups

In the case of Boolean groups, many assertions concerning large sets can be refined. For example, properties 10 and 11 of large sets are stated as follows.

Proposition 5. *The following assertions hold.*

(i) *For any thick set T in a Boolean group G with zero 0 , there exists an infinite subgroup H of G for which* $T \cup \{0\} \supset H$.
(ii) *Any set which intersects nontrivially all infinite subgroups in a Boolean group G is syndetic.*

Note that this is not so in non-Boolean groups: the set $\{n! : n \in \mathbb{N}\}$ intersects any infinite subgroup in $\mathbb{Z}$, but it is not syndetic, because the gaps between neighboring elements are not bounded. The complement of this set contains no infinite subgroups, and it is thick by Property 2 of large sets.

Another specific feature of thick sets in Boolean groups is given by the following proposition.

Proposition 6. *For any thick set T in a countable Boolean group G with zero 0, there exists a set* $A \subset G$ *such that* $T \cup \{0\} = A \bigtriangleup A$ *(and* $A \bigtriangleup A \bigtriangleup A \bigtriangleup A = G$ *by Theorem 3 (iv)).*

Proposition 6 is an immediate corollary of Lemma 4.3 in [3], which says that any thick set in a countable Abelian group equals $\Delta_I((g_n)_{n=1}^\infty)$ for some sequence $(g_n)_{n=1}^\infty$.

In view of Example 4, we cannot assert that the set A in this proposition is large (in whatever sense), even for the largest (3-vast) nontrivial thick sets T.

The following statement can be considered as a partial analogue of Propositions 5 and 6 for Δ^*-(in particular, vast) sets in Boolean groups.

Theorem 5. *For any Δ^*-set A in a Boolean group G with zero $\mathbf{0}$, there exists a $B \subset G$ with $|B| = |A|$ such that $B \mathbin{\triangle} B \subset A \cup \{\mathbf{0}\}$.*

Proof. First, note that $|A| = |G|$. Any Boolean group is algebraically free; therefore, we can assume that $G = B(X)$ for a set X with $|X| = |A|$. Let

$$A_2 = A \cap B_{=2}(X) = \{\{x,y\} = x \mathbin{\triangle} y \in A : x, y \in X\}$$

be the intersection of A with the set of elements of length 2. We have $|A_2| = |X|$, because A must intersect nontrivially each countable set of the form $Y \mathbin{\triangle} Y$ for $Y \subset X$. Consider the coloring $c \colon [X]^2 \to \{0,1\}$ defined by

$$c(\{x,y\}) = \begin{cases} 0 & \text{if } \{x,y\} \in A_2, \\ 1 & \text{otherwise.} \end{cases}$$

According to the well-known Erdős–Dushnik–Miller theorem $\kappa \to (\kappa, \aleph_0)^2$ (see, e.g., [19]), there exists either an infinite set $Y \subset X$ for which $[Y]^2 \cap A_2 = \varnothing$ or a set $Y \subset X$ of cardinality $|X|$ for which $[Y]^2 \subset A_2$. The former case cannot occur, because $[Y]^2 = Y \mathbin{\triangle} Y$ in $B(X)$, $[Y]^2 \subset B_{=2}(X)$, and A is a Δ^*-set. Thus, the latter case occurs, and we set $B = Y$. $\square$

We have already distinguished between vast sets and translates of syndetic sets in Boolean groups (see Example 2). For completeness, we give the following example.

Example 5. *The countable Boolean group $B(\mathbb{Z})$ contains an IP*-set (see Property 11 of large sets) which is not a Δ^*-set. An example of such a set is constructed from the corresponding example in $\mathbb{Z}$ (see [15] p. 177) in precisely the same way as Example 2.*

8. Large Sets in Free Boolean Topological Groups

As shown in Section 5, given any Boolean group G, the filter of vast sets in G cannot be the filter of neighborhoods of zero for a group topology, because not all vast and even 3-vast sets are neighborhoods of zero in the Bohr topology. Moreover, if we fix any basis X in G, so that $G = B(X)$, then not all traces of 3-vast sets on the set $B_{=2}(X)$ of two-letter elements contain those of Bohr open sets (see Example 4). However, there are natural group topologies on $B(X)$ such that the topologies which they induce on $B_2(X)$ contain those generated by n-vast sets. These are, e.g., topologies induced on $B(X)$ from the free Boolean topological groups $B(X_{\mathscr{F}})$ for certain filters $\mathscr{F}$ on X (see Section 1). Before proceeding to main statements, we make several general observations concerning free Boolean topological groups on filter spaces $X_{\mathscr{F}}$.

Let X be a set, and let $\mathscr{F}$ be a filter on X. The free Boolean group $B(X)$ (without topology) is embedded in $B(X_{\mathscr{F}})$ as a subgroup; we denote $B(X)$ endowed with the topology induced from $B(X_{\mathscr{F}})$ by $B^{\mathrm{i}}(X)$ and use $B(X)$ to denote the abstract free Boolean group on X (without topology). Although X is discrete in $X_{\mathscr{F}}$, $B^{\mathrm{i}}(X)$ and even $B^{\mathrm{i}}_2(X)$ are not discrete: any neighborhood of zero must contain all elements $x \mathbin{\triangle} y$ with $x, y \in A$ for some $A \in \mathscr{F}$. However, the set $B^{\mathrm{i}}_{=2}(X)$ is discrete. Indeed, for any $g \in B_{=2}(X)$ and any $A \in \mathscr{F}$, the set $g \mathbin{\triangle} \langle A \mathbin{\triangle} * \rangle$ is a neighborhood of g in $B(X_{\mathscr{F}})$, and if none of the two letters of g belongs to A, then this neighborhood contains no elements of $B_{=2}(X)$ other than g. Note also that the Graev free Boolean group $B^G(X_{\mathscr{F}})$ (with zero $*$) treated as a set, that is, $([X]^{<\omega} \setminus \{\varnothing\}) \cup \{\{*\}\}$, is a subset of $B(X_{\mathscr{F}}) = [X \cup \{*\}]^{<\omega}$. Moreover, $B^G_{=n}(X_{\mathscr{F}}) = B_{=n}(X)$ for each $n > 0$, and a set $C \subset B(X)$ is k-vast in $B(X)$ if and only if $(C \setminus \{\varnothing\}) \cup \{\{*\}\}$ is k-vast in $B^G(X_{\mathscr{F}})$ (It is also easy to see that any such set C is $\leq 2k$-vast in $B(X_{\mathscr{F}})$). These observations imply the following proposition, which helps to better understand the meaning of the main theorems.

Proposition 7. *Let* $n \in \mathbb{N}$, *and let* $\mathscr{F}$ *be a filter on an infinite set* X.

(i) *Suppose that* $Y \subset B_{=2n}(X)$ *and* U *is a neighborhood of zero in the free group topology of* $B(X_{\mathscr{F}})$. *Then* Y *is dense in* $U \cap B_{=2n}(X_{\mathscr{F}})$ *(in the topology of* $B(X_{\mathscr{F}})$*) if and only if* $Y = U \cap B_{=2n}(X)$.

(ii) *A set* $Y \subset B_2(X)$ *contains the trace on* $B_2(X)$ *of a neighborhood of zero in* $B(X_{\mathscr{F}})$ *if and only if* Y *is dense in the trace on* $B_2(X)$ *of such a neighborhood.*

(iii) *A set* $Y \subset B_{=2n}(X) = B^G_{=2n}(X_{\mathscr{F}})$ *contains the trace on* $B_{=2n}(X)$ *of a neighborhood of zero in* $B(X_{\mathscr{F}})$ *if and only if* $Y = U \cap B^G_{=2n}(X_{\mathscr{F}})$ *for a neighborhood* U *of zero in* $B^G(X_{\mathscr{F}})$.

(iv) *Let* $B^G_{\text{even}}(X_{\mathscr{F}})$ *denote the subgroup of* $B^G(X_{\mathscr{F}})$ *consisting of all elements of even length. This subgroup is naturally topologically isomorphic to* $B^G([X]^2_{\mathscr{F}'})$, *where* $\mathscr{F}'$ *is the filter on* $[X]^2$ *generated by sets of the form* $[A]^2$ *for* $A \in \mathscr{F}$. *A set* $Y \subset B(X)$ *is a neighborhood of zero in* $B^i(X)$ *if and only if* $((Y \setminus \{\varnothing\}) \cup \{\{*\}\}) \cap B^G_{\text{even}}(X_{\mathscr{F}})$ *is a neighborhood of zero in* $B^G_{\text{even}}(X_{\mathscr{F}})$.

Proof. (i) First, note that $U \cap B_{=2n}(X)$ is dense in $U \cap B_{=2n}(X_{\mathscr{F}})$. Indeed, each $g \in U \cap (B_{=2n}(X_{\mathscr{F}}) \setminus B_{=2n}(X)$ has the form $* \vartriangle x_1 \vartriangle x_2 \vartriangle \cdots \vartriangle x_{2n-1}$, where $x_i \in X$, and for any such g, there exists an $A \in \mathscr{F}$ such that $x \vartriangle x_1 \vartriangle x_2 \vartriangle \cdots \vartriangle x_{2n-1} \in U \cap B_{=2n}(X)$ for every $x \in A$. This proves the "if" part. Conversely, since $B^i_{=2n}(X)$ is discrete, it follows that a subset of $B_{=2n}(X)$ is dense in another subset of $B_{=2n}(X)$ in the topology of $B(X_{\mathscr{F}})$ if and only if these subsets coincide. Thus, if Y is dense in $U \cap B_{=2n}(X_{\mathscr{F}})$ (and hence in $U \cap B^i_{=2n}(X)$), then Y must coincide with $U \cap B_{=2n}(X)$.

Assertion (ii) follows from (i) and the observation that $B_2(X) = B_{=2}(X) \cup \{0\}$.

Assertions (iii) and (iv) follow directly from the descriptions of the topologies of $B(X_{\mathscr{F}})$ and $B^G(X_{\mathscr{F}})$ given in Section 1. $\quad\square$

Theorem 6. *Let* $k \in \mathbb{N}$, *and let* $\mathscr{F}$ *be a filter on an infinite set* X. *Then the following assertions hold.*

(i) *For* $k \neq 4$, *the trace of any k-vast subset of* $B(X)$ *on* $B_2(X) \subset B_2(X_{\mathscr{F}})$ *contains that of a neighborhood of zero in the free group topology of* $B(X_{\mathscr{F}})$ *if and only if* $\mathscr{F}$ *is a k-arrow filter.*

(ii) *If the trace of any 4-vast set on* $B_2(X)$ *contains that of a neighborhood of zero in the free group topology of* $B(X_{\mathscr{F}})$, *then* $\mathscr{F}$ *is a 4-arrow filter, and if* $\mathscr{F}$ *is a 4-arrow filter, then the trace of any 3-vast set on* $B_2(X_{\mathscr{F}})$ *contains that of a neighborhood of zero in the free group topology of* $B(X_{\mathscr{F}})$.

(iii) *The trace of any ω-vast set on* $B_2(X)$ *contains that of a neighborhood of zero in the free group topology of* $B(X_{\mathscr{F}})$ *if and only if* $\mathscr{F}$ *is an ω-arrow ultrafilter.*

The proof of this theorem uses the following lemma.

Lemma 1. *The following assertions hold.*

(i) *If* $k \neq 4$, $w_1, \ldots, w_k \in B(X)$, *and* $w_i \vartriangle w_j \in B_{=2}(X)$ *for any* $i < j \leq k$, *then there exist* $x_1, \ldots, x_k \in X$ *such that* $w_i \vartriangle w_j = x_i \vartriangle x_j$ *for any* $i < j \leq k$.

(ii) *If* $k = 4$, $w_1, w_2, w_3, w_4 \in B(X)$, *and* $w_i \vartriangle w_j \in B_{=2}(X)$ *for any* $i < j \leq 4$, *then there exist either*

    (a) $x_1, x_2, x_3, x_4 \in X$ *such that* $w_i \vartriangle w_j = x_i \vartriangle x_j$ *for any* $i < j \leq 4$ *or*
    (b) $x_1, x_2, x_3 \in X$ *such that*

$$w_1 \vartriangle w_4 = w_2 \vartriangle w_3 = x_2 \vartriangle x_3,$$
$$w_2 \vartriangle w_4 = w_1 \vartriangle w_3 = x_1 \vartriangle x_3,$$
$$w_3 \vartriangle w_4 = w_1 \vartriangle w_3 = x_1 \vartriangle x_3.$$

(iii) *If* $w_1, w_2, \cdots \in B(X)$ *and* $w_i \vartriangle w_j \in B_2(X)$ *for any* $i < j$, *then there exist* $x_1, x_2, \cdots \in X$ *such that* $w_i \vartriangle w_j = x_i \vartriangle x_j$ *for any* $i < j$.

Proof. We prove the lemma by induction on k. There is nothing to prove for $k = 1$, and for $k = 2$, assertion (i) obviously holds.

Suppose that $k = 3$. For some $y_1, y_2, y_3, y_4 \in X$, we have $w_1 \triangle w_2 = y_1 \triangle y_2$ and $w_2 \triangle w_3 = y_3 \triangle y_4$. Since $w_1 \triangle w_3 = w_1 \triangle w_2 \triangle w_2 \triangle w_3 \in B_{=2}(X)$, it follows that either $y_1 = y_3$, $y_1 = y_4$, $y_2 = y_3$, or $y_2 = y_4$. If $y_1 = y_3$, then $w_1 \triangle w_3 = y_2 \triangle y_4$ and $w_2 \triangle w_3 = y_1 \triangle y_4$, so that we can set $x_1 = y_2$, $x_2 = y_1$, and $x_3 = y_4$. If $y_2 = y_3$, then $w_1 \triangle w_3 = y_1 \triangle y_4$ and $w_2 \triangle w_3 = y_2 \triangle y_4$, and we set $x_1 = y_1$, $x_2 = y_1$, and $x_3 = y_4$. The remaining cases are treated similarly.

Suppose that $k = 4$ and let $x_1, x_2, x_3 \in X$ be such that $w_i \triangle w_j = x_i \triangle x_j$ for $i = 1,2,3$. There exist $y, z \in X$ for which $w_1 \triangle w_4 = y \triangle z$. We have $w_2 \triangle w_4 = w_1 \triangle w_2 \triangle w_1 \triangle w_4 = x_1 \triangle x_2 \triangle y \triangle z \in B_2(X)$. Therefore, either $x_1 = y$, $x_2 = y$, $x_1 = z$, or $x_2 = z$.

If $x_1 = y$ or $x_1 = z$, then the condition in (ii) (a) holds for $x_4 = z$ in the former case and $x_4 = y$ in the latter.

Suppose that $x_1 \neq y$ and $x_1 \neq z$. Then $x_2 = y$ or $x_2 = z$. Let $x_2 = y$. Then $w_1 \triangle w_4 = x_2 \triangle z$, and we have $w_3 \triangle w_4 = w_1 \triangle w_3 \triangle w_1 \triangle w_4 = x_1 \triangle x_3 \triangle x_2 \triangle z \in B_2(X)$, whence $x_3 = z$ (because $x_1, x_2 \neq z$), so that $w_1 \triangle w_4 = x_2 \triangle x_3 = w_2 \triangle w_3$, $w_2 \triangle w_4 = w_1 \triangle w_2 \triangle w_1 \triangle w_4 = x_1 \triangle x_3 = w_1 \triangle w_3$, and $w_3 \triangle w_4 = x_1 \triangle x_3 = w_1 \triangle w_3$, i.e., assertion (ii) (b) holds. The case $x_2 = z$ is similar. Note for what follows that, in both cases $x_2 = y$ and $x_2 = z$, we have $w_4 = w_1 \triangle w_2 \triangle w_3$.

Let $k > 4$. Consider w_1, w_2, w_3, and w_4. Let $x_1, x_2, x_3 \in X$ be such that $w_i \triangle w_j = x_i \triangle x_j$ for $i = 1,2,3$. As previously, there exist $y, z \in X$ for which $w_1 \triangle w_4 = y \triangle z$ and either $x_1 = y$, $x_2 = y$, $x_1 = z$, or $x_2 = z$.

Suppose that $x_1 \neq y$ and $x_1 \neq z$; then $w_4 = w_1 \triangle w_2 \triangle w_3$. In this case, we consider w_5 instead of w_4. Again, there exist $y', z' \in X$ for which $w_1 \triangle w_5 = y \triangle z$ and either $x_1 = y'$, $x_2 = y'$, $x_1 = z'$, or $x_2 = z'$. Since $w_5 \neq w_4$, it follows that $w_5 \neq w_1 \triangle w_2 \triangle w_3$, and we have $x_1 = y'$ or $x_1 = z'$. In the former case, we set $x_5 = z'$ and in the latter, $x_5 = y'$. Consider again w_4; recall that $w_1 \triangle w_4 = y \triangle z$. We have $w_i \triangle w_4 = w_1 \triangle w_i \triangle w_1 \triangle w_4 = x_1 \triangle x_i \triangle y \triangle z \in B_{=2}(X)$ for $i \in \{2,3,5\}$. Since $x_2 \neq x_5$ and $x_3 \neq x_5$, it follows that $x_1 = y$, which contradicts the assumption.

Thus, $x_1 = y$ or $x_1 = z$. As above, we set $x_4 = z$ in the former case and $x_4 = y$ in the latter; then the condition in (ii) (a) holds.

Suppose that we have already found the required $x_1, \ldots, x_{k-1} \in X$ for $w_1, \ldots, w_{k-1}$. There exist $y, z \in X$ for which $w_1 \triangle w_k = y \triangle z$. We have $w_i \triangle w_k = w_1 \triangle w_i \triangle w_1 \triangle w_k = x_1 \triangle x_i \triangle y \triangle z \in B_{=2}(X)$ for $i \leq k - 1$. If $x_1 \neq y$ and $x_1 \neq z$, then we have $x_i \in \{y, z\}$ for $2 \leq i \leq k - 1$, which is impossible, because $k > 4$. Thus, either $x_1 = y$ or $x_1 = z$. In the former case, we set $x_k = z$ and in the latter, $x_k = y$. Then $w_1 \triangle w_k = x_1 \triangle w_k$ and, for any $i \leq k - 1$, $w_i \triangle w_k = w_1 \triangle w_i \triangle w_1 \triangle w_k = x_1 \triangle x_i \triangle x_1 \triangle x_k = x_i \triangle x_k$.

The infinite case is proved by the same inductive argument. □

Proof of Theorem 6. (i) Suppose that $\mathscr{F}$ is a k-arrow filter on X. Let C be a k-vast set in $B(X)$. Consider the 2-coloring of $[X]^2$ defined by

$$c(\{x,y\}) = \begin{cases} 0 & \text{if } \{x,y\} = x \triangle y \in C, \\ 1 & \text{otherwise.} \end{cases}$$

Since $\mathscr{F}$ is k-arrow, there exists either an $A \in \mathscr{F}$ for which $c([A]^2) = \{0\}$ and hence $[A]^2 \subset C \cap B_2(X_{\mathscr{F}})$ or a k-element set $F \subset X$ for which $c([F]^2) = \{1\}$ and hence $[F]^2 \cap C = [F]^2 \cap C \cap B_{=2}(X_{\mathscr{F}}) = \varnothing$. The latter case cannot occur, because C is k-vast. Therefore, $C \cap B_2(X_{\mathscr{F}})$ contains the trace $[A]^2 \cup \{0\} = ((A \cup \{*\}) \triangle (A \cup \{*\}))$ of the subgroup $\langle A \cup \{*\}\rangle$, which is an open neighborhood of zero in $B(X_{\mathscr{F}})$.

Now suppose that $k \neq 4$ and the trace of each k-vast set on $B_2(X)$ contains the trace on $B_2(X)$ of a neighborhood of zero in $B(X_{\mathscr{F}})$, i.e., a set of the form $A \triangle A$ for some $A \in \mathscr{F}$. Let us show that $\mathscr{F}$ is k-arrow. Given any $c \colon [X]^2 \to \{0,1\}$, we set

$$C = \{x \triangle y : c(\{x,y\}) = 1\} \quad \text{and} \quad C' = B(X_{\mathscr{F}}) \setminus C.$$

If C' is not k-vast, then there exist $w_1, \ldots, w_k \in B(X)$ such that $w_i \bigtriangleup w_j \in C$ for $i < j \leq k$. By Lemma 1 (i) we can find $x_1, \ldots, x_k \in X$ such that $x_i \bigtriangleup x_j \in C$ (and hence $x_i \neq *$) for $i < j \leq k$. This means that, for $F = \{x_1, \ldots, x_k\}$, we have $c([F]^2) = \{1\}$. If C' is k-vast, then, by assumption, there exists an $A \in \mathscr{F}$ for which $A \bigtriangleup A \setminus \{0\} \subset C' \cap B_2(X) = C$, which means that $c([A]^2) = \{0\}$.

The same argument proves (ii); the only difference is that assertion (ii) of Lemma 1 is used instead of (i).

The proof of (iii) is similar. $\quad\square$

Let $R_r(s)$ denote the least number n such that, for any r-coloring $c\colon [X]^2 \to Y$, where $|X| \geq n$ and $|Y| = r$, there exists an s-element c-homogeneous set. By the finite Ramsey theorem, such a number exists for any positive integers r and s.

Theorem 7. *There exists a positive integer N (namely, $N = R_{36}(R_6(3)) + 1$) such that, for any uniform ultrafilter $\mathscr{U}$ on a set X of infinite cardinality κ, the following conditions are equivalent:*

(i) *the trace of any N-vast subset of $B(X)$ on $B_4(X) \subset B_4(X_{\mathscr{U}})$ contains that of a neighborhood of zero in the free group topology of $B(X_{\mathscr{U}})$;*

(ii) *all κ-vast sets in $B(X)$ are neighborhoods of zero in the topology induced from the free topological group $B(X_{\mathscr{U}})$;*

(iii) *$\mathscr{U}$ is a Ramsey ultrafilter.*

Proof. Without loss of generality, we assume that $X = \kappa$.

(i) $\Rightarrow$ (iii) Suppose that N is as large as we need and the trace of each N-vast set on $B_4(\kappa_{\mathscr{U}})$ contains the trace on $B_4(\kappa)$ of a neighborhood of zero in $B(X_{\mathscr{U}})$, which, in turn, contains a set of the form $(A \bigtriangleup A \bigtriangleup A \bigtriangleup A) \cap B_{=4}(\kappa)$ for some $A \in \mathscr{U}$. Let us show that $\mathscr{U}$ is a Ramsey ultrafilter. Consider any 2-coloring $c\colon [\kappa]^2 \to \{0, 1\}$. We set

$$C = \big\{ \alpha_1 \bigtriangleup \alpha_2 \bigtriangleup \alpha_3 \bigtriangleup \alpha_4 : \alpha_i \in \kappa \text{ for } i \leq 4, \; \alpha_1 < \alpha_2 < \alpha_3 < \alpha_4,$$
$$c(\{\alpha_1, \alpha_2\}) \neq c(\{\alpha_3, \alpha_4\}), \; c(\{\alpha_1, \alpha_3\}) \neq c(\{\alpha_2, \alpha_4\}),$$
$$c(\{\alpha_1, \alpha_4\}) \neq c(\{\alpha_2, \alpha_3\}) \big\}$$

and

$$C' = B(X) \setminus C.$$

If C' is not N-vast, then there exist $w_1, \ldots, w_N \in B(\kappa)$ such that $w_i \bigtriangleup w_j \in C$ for $i < j \leq N$. We can assume that $w_N = 0$ (otherwise, we translate all w_i by w_N). Then $w_i \in C \subset B_4(\kappa)$, $i < N$. Let $w_i = \alpha_1^i \bigtriangleup \alpha_2^i \bigtriangleup \alpha_3^i \bigtriangleup \alpha_4^i$ for $i < N$ and consider the 36-coloring of all pairs $\{w_i, w_j\}$, $i < j < N$, defined as follows. Since $w_i \bigtriangleup w_j$ is a four-letter element, it follows that $w_i \bigtriangleup w_j = \beta_1 \bigtriangleup \beta_2 \bigtriangleup \beta_3 \bigtriangleup \beta_4$, where $\beta_i \in \kappa$. Two letters among $\beta_1, \beta_2, \beta_3, \beta_4$ (say β_1 and β_2) occur in w_i and the remaining two (β_3 and β_4) occur in w_j. We assume that $\beta_1 < \beta_2$ and $\beta_3 < \beta_4$. Let us denote the numbers of the letters β_1 and β_2 in w_i (recall that the letters in w_i are numbered in increasing order) by i' and i'', respectively, and the numbers of the letters β_3 and β_4 in w_j by j' and j''. To the pair $\{w_i, w_j\}$ we assign the quadruple (i', i'', j', j''). The number of all possible quadruples is 36, so that this assignment is a 36-coloring. We choose $N \geq R_{36}(N') + 1$ for N' as large as we need. Then there exist two pairs i'_0, i''_0 and j'_0, j''_0 and N' elements w_{i_n}, where $n \leq N'$ and $i_s < i_t$ for $s < t$, such that $i' = i'_0$, $i'' = i''_0$, $j' = j'_0$, and $j'' = j''_0$ for any pair $\{w_i, w_j\}$ with $i, j \in \{i_1, \ldots, i_{N'}\}$ and $i < j$. Clearly, if $N' \geq 3$, then we also have $j'_0 = i'_0$ and $j''_0 = i''_0$. In the same manner, we can fix the position of the letters coming from w_i and w_j in the sum $w_i \bigtriangleup w_j$: to each pair $\{w_{i_s}, w_{i_t}\}$, $s, t \in \{1, \ldots, N'\}$, $s < t$, we assign the numbers of the i'_0th and i''_0th letters of w_{i_s} in $w_{i_s} \bigtriangleup w_{i_t}$ (recall that the letters are numbered in increasing order); the positions of the letters of w_{i_t} in $w_{i_s} \bigtriangleup w_{i_t}$ are then determined automatically. There are six possible arrangements: 1,2, 1,3, 1,4, 2,3, 2,4, and 3,4. Thus, we have a 6-coloring of the symmetric square of the N'-element set $\{w_{i_1}, \ldots, w_{i_{N'}}\}$, and if $N' \geq R_6(3)$ (which we assume), then there exists a 3-element set $\{w_k, w_l, w_m\}$ homogeneous with respect to this coloring, i.e., such that all pairs of elements from this set are assigned the same color. For definiteness, suppose that this is the color 1, 2; suppose also that $i_0' = 1$, $i_0'' = 2$, $k < l < m$, and $w_t = a_1^t ▵ a_2^t ▵ a_3^t ▵ a_4^t$ for $t = k, l, m$. Then $w_k, w_l, w_m \in C$, $w_k ▵ w_l = a_1^k ▵ a_2^k ▵ a_1^l ▵ a_2^l \in C$, $w_l ▵ w_m = a_1^l ▵ a_2^l ▵ a_1^m ▵ a_2^m \in C$, and $w_k ▵ w_m = a_1^k ▵ a_2^k ▵ a_1^m ▵ a_2^m \in C$. By the definition of C we have $c(a_1^k ▵ a_2^k) \neq c(a_1^l ▵ a_2^l)$, $c(a_1^l ▵ a_2^l) \neq c(a_1^m ▵ a_2^m)$, and $c(a_1^k ▵ a_2^k) \neq c(a_1^m ▵ a_2^m)$, which is impossible, because c takes only two values. The cases of other colors and other numbers i_0' and i_0'' are treated in a similar way.

Thus, C' is N-vast and, therefore, contains $(A ▵ A ▵ A ▵ A) \cap B_4(\kappa)$ for some $A \in \mathscr{U}$. Take any $\alpha \in A$ and consider the sets $A' = \{\beta > \alpha : c(\{\alpha, \beta\}) = \{0\}\}$ and $A'' = \{\beta > \alpha : c(\{\alpha, \beta\}) = \{1\}\}$. One of these sets belongs to $\mathscr{U}$, because $\mathscr{U}$ is uniform. For definiteness, suppose that this is A'. By Theorem 6 $\mathscr{U}$ is 3-arrow. Hence there exists either an $A'' \subset A'$ for which $c([A'']^2) = \{0\}$ or $\beta, \gamma, \delta \in A'$, $\beta < \gamma < \delta$, for which $c([\{\beta, \gamma, \delta\}]^2) = \{1\}$. In the former case, we are done. In the latter case, we have $\alpha, \beta, \gamma, \delta \in A$, $\alpha < \beta < \gamma < \delta$, $c(\{\beta, \gamma\}) = c(\{\gamma, \delta\}) = c(\{\beta, \delta\}) = 1$, and $c(\{\alpha, \beta\}) = c(\{\alpha, \gamma\}) = c(\{\alpha, \delta\}) = 0$ (by the definition of A'). Therefore, $\alpha ▵ \beta ▵ \gamma ▵ \delta \in C$, which contradicts the definition of A.

(iii) $\Rightarrow$ (ii) Suppose that $\mathscr{U}$ is a Ramsey ultrafilter on X and C is a κ-vast set in $B(X)$. Take any $n \in \mathbb{N}$ and consider the coloring $c: [X]^{2n} \to \{0, 1\}$ defined by

$$c(\{x_1, \ldots, x_{2n}\}) = \begin{cases} 0 & \text{if } \{x_1, \ldots, x_{2n}\} = x_1 ▵ \cdots ▵ x_{2n} \in C, \\ 1 & \text{otherwise.} \end{cases}$$

Since $\mathscr{U}$ is Ramsey, there exists either a set $A_n \in \mathscr{U}$ for which $[A]^{2n} \subset C$ or a set $Y \subset X$ of cardinality κ for which $[Y]^{2n} \cap C = \varnothing$. In the latter case, for $Z = [Y]^n \subset B(X)$, we have $(Z ▵ Z) \cap C \subset \{0\}$, which contradicts C being κ-vast. Hence the former case occurs, and $C \cap B_{2n}(X)$ contains the trace $[A_n]^{2n} \cap B_{=2n}(X)$ of the open subgroup $\langle (A_n \cup \{*\}) ▵ (A_n \cup \{*\}) \rangle$ of $B(X_{\mathscr{F}})$.

Thus, for each $n \in \mathbb{N}$, we have found $A_1, A_2, \ldots, A_n \in \mathscr{F}$ such that $[A_i]^{2i} \cap B_{=2i}(X) \subset C$. Let $A = \bigcap_{i \leq n} A_i$. Then $A \in \mathscr{U}$ and $[A]^{2i} \cap B_{=2i}(X) \subset C$ for all $i \leq n$. Hence $C \cap B_{2n}(X)$ contains the trace on $B_{2n}(X)$ of the open subgroup $\langle (A \cup \{*\}) ▵ (A \cup \{*\}) \rangle$ of $B(X_{\mathscr{U}})$ (recall that $0 \in C$). This means that, for each n, $C \cap B_{2n}(X)$ is a neighborhood of zero in the topology induced from $B(X_{\mathscr{U}})$.

If $\kappa = \omega$, then $B(X_{\mathscr{U}})$ has the inductive limit topology with respect to the decomposition $B(X_{\mathscr{U}}) = \bigcup_{n \in \omega} B_n(X_{\mathscr{F}})$, because $\mathscr{F}$ is Ramsey (see [4]). Therefore, in this case, $C \cap B(X)$ is a neighborhood of zero in the induced topology.

If $\kappa > \omega$, then the ultrafilter $\mathscr{U}$ is countably complete ([8] Lemma 9.5 and Theorem 9.6), i.e., any countable intersection of elements of $\mathscr{U}$ belongs to $\mathscr{U}$. Hence $A = \bigcap_{n \in \mathbb{N}} A_n \in \mathscr{U}$, and $\langle (A \cup \{*\}) ▵ (A \cup \{*\}) \rangle \cap \left(\bigcup_{n \in \omega} B_{2n}(X) \right) \subset C$. Thus, $C \cap B(X)$ is a neighborhood of zero in the induced topology in this case, too.

The implication (ii) $\Rightarrow$ (i) is obvious. □

Theorem 6 has the following purely algebraic corollary.

Corollary 2 ($\mathfrak{p} = \mathfrak{c}$). *Any Boolean group contains ω-vast sets which are not vast and Δ^*-sets which are Δ_k^*-sets for no k.*

Proof. Theorem 4.10 of [9] asserts that if $\mathfrak{p} = \mathfrak{c}$, then there exists an ultrafilter $\mathscr{U}$ on ω which is k-arrow for all $k \in \mathbb{N}$ but not Ramsey and, therefore, not ω-arrow ([9] Theorem 2.1). By Theorem 6 the traces of all vast sets on $B_2(\omega)$ contain those of neighborhoods of zero in $B(\omega_{\mathscr{U}})$, and there exist ω-vast sets whose traces do not. This proves the required assertion for the countable Boolean group. The case of a group $B(X)$ of uncountable cardinality κ reduces to the countable case by representing $B(X)$ as $B(\kappa) = B(\omega) \times B(\kappa)$; it suffices to note that a set of the form $C \times B(\kappa)$, where $C \subset B(\omega)$, is λ-vast in $B(\omega) \times B(\kappa)$ for $\lambda \leq \omega$ if and only if so is C in $B(\omega)$. □

The author is unaware of where there exist ZFC examples of such sets in any groups.

Acknowledgments: The author thanks Evgenii Reznichenko and Anton Klyachko for useful discussions. The author is very grateful to the referees for just criticism, which helped to substantially improve the exposition. This work was financially supported by the Russian Foundation for Basic Research (Project No. 17-51-18051).

Conflicts of Interest: The authors declare no conflict of interest.

References

1. Gottschalk,W.H.; Hedlund, G.A. *Topological Dynamics*; American Mathematical Society: Providence, RI, USA, 1955.
2. Reznichenko, E.; Sipacheva, O. Discrete Subsets in Topological Groups and Countable Extremally Disconnected Groups. *arXiv* **2016**, arXiv:1608.03546.
3. Bergelson, V.; Hindman, N.; McCutcheon, R. Notions of size and combinatorial properties of quotient sets in semigroups. *Topol. Proc.* **1998**, *23*, 23–60.
4. Sipacheva, O. Free Boolean topological groups. *Axioms* **2015**, *4*, 492–517.
5. Mokobodzki, G. *Ultrafiltres Rapides sur* ℕ. *Construction D'une Densité Relative de Deux Potentiels Comparables*; Théorie du potentiel; Séminaire Brelot-Choquet-Deny; Secrétariat Mathématique: Paris, France, 1967–1968; Volume 12, pp. 1–22.
6. Mathias, A.R.D. A remark on rare filters. In *Infinite and Finite Sets*; Colloquia Mathematica Societatis János Bolyai: Amsterdam, The Netherlands, 1975; Volume 3, pp. 1095–1097.
7. Miller, A.W. There are no *q*-points in Laver's model for the borel conjecture. *Proc. Am. Math. Soc.* **1980**, *78*, 103–106.
8. Comfort, W.W.; Negrepontis, S. *The Theory of Ultrafilters*; Springer: Berlin, Germany, 1974.
9. Baumgartner, J.E.; Taylor, A.D. Partition theorems and ultrafilters. *Trans. Am. Math. Soc.* **1978**, *241*, 283–309.
10. Shelah, S. *Proper and Improper Forcing*; Springer: Berlin, Germany, 1998.
11. Van Douwen, E.K. The integers and topology. In *Handbook of Set-Theoretic Topology*; Elsevier: Amsterdam, The Netherlands, 1984; pp. 111–167.
12. Bergelson, V.; Hindman, N. Partition regular structures contained in large sets are abundant. *J. Comb. Theory* **2001**, *93*, 18–36.
13. Hindman, N.; Strauss, D. *Algebra in the Stone—Čech Compactification*, 2nd ed.; De Gruyter: Berlin, Germany; Boston, MA, USA, 2012.
14. Bergelson, V.; Furstenberg, H.; Weiss, B. Piecewise-Bohr sets of integers and combinatorial number theory. In *Topics in Discrete Mathematics*; Klazar, M., Kratochvíl, J., Loebl, M., Matoušek, J., Valtr, P., Thomas, R., Eds.; Springer: Berlin/Heidelberg, Germany, 2006; pp. 13–37.
15. Furstenberg, H. *Recurrence in Ergodic Theory and Combinatorial Number Theory*; Princeton University Press: Princeton, NJ, USA, 1981.
16. Griesmer, J.T. (Department of Applied Mathematics and Statistics, Colorado School of Mines, Golden, CO, USA). A Remark on Intersectivity and Difference Sets. Personal communication, 2016.
17. Ellis, R.; Keynes, H.B. Bohr compactifications and a result of Følner. *Isr. J. Math.* **1972**, *12*, 314–330.
18. Beiglböck, M.; Bergelson, V.; Alexander, F. Sumset phenomenon in countable amenable groups. *Adv. Math.* **2010**, *223*, 416–432.
19. Jech, T. *Set Theory*, 3rd millennium (revised) ed.; Springer: Berlin, Germany, 2003.

axioms

MDPI

Article

Selectively Pseudocompact Groups without Infinite Separable Pseudocompact Subsets †

Dmitri Shakhmatov [1,*] **and Víctor Hugo Yañez** [2]

1   Division of Mathematics, Physics and Earth Sciences, Graduate School of Science and Engineering, Ehime University, Matsuyama 790-8577, Japan
2   Doctor's Course, Graduate School of Science and Engineering, Ehime University, Matsuyama 790-8577, Japan; victor_yanez@comunidad.unam.mx
*   Correspondence: dmitri.shakhmatov@ehime-u.ac.jp; Tel.: +81-89-927-9558
†   This article is dedicated to Professor Alexander V. Arhangel'skiĭ on the occasion of his 80th birthday.

Received: 31 July 2018; Accepted: 5 November 2018; Published: 16 November 2018

Abstract: We give a "naive" (i.e., using no additional set-theoretic assumptions beyond ZFC, the Zermelo-Fraenkel axioms of set theory augmented by the Axiom of Choice) example of a Boolean topological group G without infinite separable pseudocompact subsets having the following "selective" compactness property: For each free ultrafilter p on the set $\mathbb{N}$ of natural numbers and every sequence (U_n) of non-empty open subsets of G, one can choose a point $x_n \in U_n$ for all $n \in \mathbb{N}$ in such a way that the resulting sequence (x_n) has a p-limit in G; that is, $\{n \in \mathbb{N} : x_n \in V\} \in p$ for every neighbourhood V of x in G. In particular, G is selectively pseudocompact (strongly pseudocompact) but not selectively sequentially pseudocompact. This answers a question of Dorantes-Aldama and the first listed author. The group G above is not pseudo-ω-bounded either. Furthermore, we show that the free precompact Boolean group of a topological sum $\bigoplus_{i \in I} X_i$, where each space X_i is either maximal or discrete, contains no infinite separable pseudocompact subsets.

Keywords: pseudocompact; strongly pseudocompact; p-compact; selectively sequentially pseudocompact; pseudo-ω-bounded; non-trivial convergent sequence; separable; free precompact Boolean group; reflexive group; maximal space; ultrafilter space

MSC: Primary: 22A05; Secondary: 54A20, 54D30, 54H11

All topological spaces considered in this paper are assumed to be Tychonoff and all topological groups are assumed to be Hausdorff (and thus Tychonoff as well).

As usual, $\mathbb{N}$ denotes the set of natural numbers, and ω denotes the first infinite cardinal. We freely identify $\mathbb{N}$ with ω. The symbol $\beta\mathbb{N}$ denotes the Stone-Čech compactification of $\mathbb{N}$. Recall that $\beta\mathbb{N} \setminus \mathbb{N}$ can be identified with the set of all free ultrafilters on $\mathbb{N}$. For sets X and Y, the symbol Y^X denotes the set of all functions from X to Y.

A group, of which each element has order 2, is called a *Boolean* group. Every Boolean group is abelian, so $x + x = 0$ holds for each element x of a Boolean group. We use $\mathbb{Z}_2$ to denote the unique (Boolean) group with two elements.

1. Definitions

Let p be a free ultrafilter on $\mathbb{N}$. Recall that a point x of a topological space X is a *p-limit* of a sequence $\{x_n : n \in \mathbb{N}\}$ of points of X provided that $\{n \in \mathbb{N} : x_n \in V\} \in p$ for every neighbourhood V of x in X [1].

The next notion is due to Angoa, Ortiz-Castillo, and Tamariz-Mascarúa [2,3].

Definition 1. *Let p be a free ultrafilter on* $\mathbb{N}$. *A space X is* strongly *p*-pseudocompact *if it has the following property: For every sequence* $\{U_n : n \in \mathbb{N}\}$ *of non-empty open subsets of X, one can choose a point* $x_n \in U_n$ *for all* $n \in \mathbb{N}$ *in such a way that the resulting sequence* $\{x_n : n \in \mathbb{N}\}$ *has a p-limit in X.*

We shall also consider a weaker property.

Definition 2. *A space X is* selectively pseudocompact *(called also* strongly pseudocompact*) provided that, for every sequence* $\{U_n : n \in \mathbb{N}\}$ *of non-empty open subsets of X, one can choose a point* $x_n \in U_n$ *for all* $n \in \mathbb{N}$ *in such a way that the resulting sequence* $\{x_n : n \in \mathbb{N}\}$ *has a p-limit in X for some free ultrafilter p on* $\mathbb{N}$ *(depending on the sequence* $\{U_n : n \in \mathbb{N}\}$ *in question).*

This notion was introduced by García-Ferreira and Ortiz-Castillo [4] under the name "strongly pseudocompact." Dorantes-Aldama and the first listed author gave a list of equivalent descriptions of this property in ([5], Theorem 2.1) and proposed an alternative name for it, calling a space with this property "selectively pseudocompact" ([5], Definition 2.2). This terminology was later adopted in [6].

Clearly, strongly *p*-pseudocompact spaces are selectively pseudocompact (strongly pseudocompact). The following notion is due to Dorantes-Aldama and the first listed author ([5], Definition 2.3).

Definition 3. *A space X is* selectively sequentially pseudocompact *provided that, for every sequence* $\{U_n : n \in \mathbb{N}\}$ *of non-empty open subsets of X, one can choose a point* $x_n \in U_n$ *for all* $n \in \mathbb{N}$ *in such a way that the resulting sequence* $\{x_n : n \in \mathbb{N}\}$ *has a convergent subsequence.*

Selectively sequentially pseudocompact spaces are selectively pseudocompact (strongly pseudocompact), while the converse does not hold in general [5].

When considering the property from Definition 1 for multiple ultrafilters *p* simultaneously, one could obtain two natural versions as follows:

Definition 4. *Let P be a non-empty subset of* $\beta\mathbb{N} \setminus \mathbb{N}$. *A space X is*

(i) strongly *P*-bounded *provided that, for every sequence* $\{U_n : n \in \mathbb{N}\}$ *of non-empty open subsets of X, one can choose a point* $x_n \in U_n$ *for all* $n \in \mathbb{N}$ *in such a way that the resulting sequence* $\{x_n : n \in \mathbb{N}\}$ *has a p-limit in X for every* $p \in P$;
(ii) strongly *P*-pseudocompact *provided that X is strongly p-pseudocompact for each* $p \in P$.

The notion of strong *P*-boundedness is due to Angoa, Ortiz-Castillo, and Tamariz-Mascarúa [2,3]. To the best of our knowledge, the notion from Item (ii) of Definition 4 appears to be new.

For every non-empty subset *P* of $\beta\mathbb{N} \setminus \mathbb{N}$, the implication

$$\text{strongly } P\text{-bounded} \rightarrow \text{strongly } P\text{-pseudocompact} \tag{1}$$

trivially holds. It is also clear that the larger the subset *P* of $\beta\mathbb{N} \setminus \mathbb{N}$ is, the stronger the corresponding property of strong *P*-boundedness and strong *P*-pseudocompactness is.

Remark 1. *(i) A sequence in a topological space X has a p-limit in X for every* $p \in \beta\mathbb{N} \setminus \mathbb{N}$ *if and only if its closure in X is compact [1]. Therefore, strong* $(\beta\mathbb{N} \setminus \mathbb{N})$*-boundedness of a space X is easily seen to be equivalent to the following property: For every sequence* $\{U_n : n \in \mathbb{N}\}$ *of non-empty open subsets of X, there exists a compact subset K of X which has a non-empty intersection with each* U_n. *The spaces having this property are called* pseudo-ω-bounded *in [2,3].*

(ii) Infinite strongly $(\beta\mathbb{N} \setminus \mathbb{N})$*-bounded spaces contain infinite compact subsets. Indeed, an infinite space X contains a sequence* $\{U_n : n \in \mathbb{N}\}$ *of pairwise disjoint non-empty open subsets. If X is strongly* $(\beta\mathbb{N} \setminus \mathbb{N})$*-bounded, then the compact subspace K of X as in Item (i) must be infinite.*

Recall that a space X is ω-*bounded* if every countable subset of X has compact closure in X. A space is *pseudocompact* if every real-valued continuous function on it is bounded.

2. Introduction

The diagram in Figure 1 summarizes implications between notions introduced in Section 1.

The double arrow in Figure 1 denotes the implication which holds only in the class of topological groups and fails for general topological spaces, as has been shown in [5].

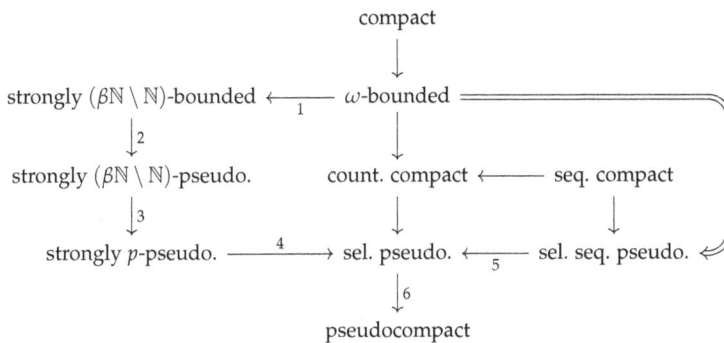

Figure 1. Implications between notions introduced in Section 1

Now we shall discuss the reversibility of arrows in Figure 1 in the class of topological groups. In Example 1, we show that Arrow 1 is not reversible. Our Corollary 2 shows that Arrow 2 is not reversible. In the text following ([7], Question 2.6), García-Ferreira and Tomita mention that there exist two free ultrafilters p and q on $\mathbb{N}$ and a topological group G which is strongly p-pseudocompact but not strongly q-pseudocompact; in particular, G is not strongly $(\beta\mathbb{N} \setminus \mathbb{N})$-pseudocompact. This shows that Arrow 3 is not reversible.

Assuming Continuum Hypothesis CH, García-Ferreira and Tomita gave an example of a selectively pseudocompact group G whose square G^2 is not selectively pseudocompact [6]. Since strong p-pseudocompactness is preserved by products [3] and implies selective pseudocompactness, G cannot be strongly p-pseudocompact for *any* free ultrafilter p on $\mathbb{N}$. This shows that Arrow 4 is not reversible under CH. The reversibility of this arrow in ZFC alone remains unclear; see Question 6.

Next, we turn our attention to Arrows 5 and 6.

García-Ferreira and Tomita in [7] gave an example demonstrating that Arrow 6 is not reversible in the class of topological groups. The authors later showed in [8] that many examples of pseudocompact groups known in the literature fail to be selectively pseudocompact, thereby establishing relative abundance of examples witnessing non-reversibility of Arrow 6 for topological groups.

Dorantes-Aldama and the first listed author gave a consistent example of a countably compact (thus, selectively pseudocompact) topological group which is not selectively sequentially pseudocompact ([5], Example 5.7), and they asked whether such an example exists in ZFC alone ([5], Question 8.3):

Question 1. (i) *Is there a ZFC example of a selectively pseudocompact (abelian) group which is not selectively sequentially pseudocompact?*

(ii) *Is there a ZFC example of a countably compact (abelian) group which is not selectively sequentially pseudocompact?*

We shall answer Item (i) of this question positively in Corollary 5, thereby showing that Arrow 5 of Figure 1 is not reversible in the class of topological groups. Moreover, an example we construct has much stronger property than mere selective pseudocompactness; see Corollary 4 (i).

Item (ii) of Question 1 remains open.

We refer the reader to [5] for examples witnessing the non-reversibility of arrows in Figure 1 without numbers assigned to them in the class of topological groups.

The paper is organized as follows. Section 3 contains our results related to Question 1. The main result here is Theorem 1. Corollary 2 in this section shows that the implication in Equation (1) is not reversible for $P = \beta \mathbb{N} \setminus \mathbb{N}$, even in the class of topological groups. Section 4 collects definitions of and background material on free Boolean groups over a set and free precompact Boolean groups of a topological space. In Section 5, we define a notion of a coherent map f and introduce a topology on its domain so that the continuity of f with respect to this topology becomes equivalent to f being coherent. Splitting maps are defined in Section 6. The notion of a coherent splitting map is used in the proof of Theorem 1. The main result in this section is Theorem 2 and its Corollary 7. In Section 7, we apply the latter to show that for every infinite subset A of the free precompact Boolean group G of an arbitrary topological sum $\bigoplus_{k \in K} X_k$, where each space X_k is either discrete or maximal, one can find a continuous group homomorphism $\varphi : G \to \mathbb{Z}_2$ such that the set $\{a \in A : \varphi(a) = z\}$ is infinite for every $z \in \mathbb{Z}_2$ (Theorem 3). This result is applied to deduce that all separable pseudocompact subsets of G as above are finite (Theorem 4). In Section 8, we discuss some connections of our results to known results in the literature. Theorem 2 is proved in Section 9, and Section 10 is devoted to the proof of Theorem 1. Open questions are listed in Section 11.

3. Results

The main goal of the paper is to prove the following theorem.

Theorem 1. *Let κ be an infinite cardinal such that $\kappa^\omega = \kappa$ and P be a non-empty subset of $\beta \mathbb{N} \setminus \mathbb{N}$ satisfying $|P| \leq \kappa$. There exists a dense strongly P-pseudocompact subgroup of $\mathbb{Z}_2^\kappa$ without infinite separable pseudocompact subsets.*

The proof of this theorem is postponed until Section 10.

Let $\mathfrak{c}$ denote the cardinality of the continuum. Applying Theorem 1 to $P = \beta \mathbb{N} \setminus \mathbb{N}$ and $\kappa = 2^\mathfrak{c}$, we obtain the following:

Corollary 1. *There exists a dense strongly $(\beta \mathbb{N} \setminus \mathbb{N})$-pseudocompact subgroup G of $\mathbb{Z}_2^{2^\mathfrak{c}}$ without infinite separable pseudocompact subsets.*

The group G in this corollary is clearly infinite. By Remark 1 (ii), infinite strongly $(\beta \mathbb{N} \setminus \mathbb{N})$-bounded spaces contain infinite compact subsets (and thus, also infinite separable pseudocompact subsets). Therefore, "strong $(\beta \mathbb{N} \setminus \mathbb{N})$-pseudocompactness" of G in Corollary 1 cannot be strengthened to its "strong $(\beta \mathbb{N} \setminus \mathbb{N})$-boundedness." By the same reason, the topological group G from Corollary 1 witnesses the validity of the following corollary, showing that Arrow 2 in Figure 1 is not reversible, even for topological groups.

Corollary 2. *A strongly $(\beta \mathbb{N} \setminus \mathbb{N})$-pseudocompact Boolean group need not be strongly $(\beta \mathbb{N} \setminus \mathbb{N})$-bounded.*

This corollary shows that the implication in Equation (1) is not reversible when $P = \beta \mathbb{N} \setminus \mathbb{N}$, even in the class of topological groups.

Given a free ultrafilter p on $\mathbb{N}$, we can apply Theorem 1 to $P = \{p\}$ and $\kappa = \mathfrak{c}$ to obtain the following:

Corollary 3. *For every free ultrafilter p on $\mathbb{N}$, there exists a dense strongly p-pseudocompact subgroup of $\mathbb{Z}_2^{\mathfrak{c}}$ without infinite separable pseudocompact subsets.*

If κ is an infinite cardinal, then every dense subset of $\mathbb{Z}_2^{\kappa}$ must be infinite. Since infinite selectively sequentially pseudocompact spaces contain non-trivial convergent sequences by ([5], Proposition 3.1) and convergent sequences are separable and pseudocompact, the topological groups from Theorem 1 and its Corollaries 1 and 3 are not selectively sequentially pseudocompact. In particular, we have the following corollary.

Corollary 4. *(i) There exists a dense strongly $(\beta\mathbb{N} \setminus \mathbb{N})$-pseudocompact subgroup of $\mathbb{Z}_2^{2^{\mathfrak{c}}}$ which is not selectively sequentially pseudocompact.*

(ii) For every free ultrafilter p on $\mathbb{N}$, there exists a dense strongly p-pseudocompact subgroup of $\mathbb{Z}_2^{\mathfrak{c}}$ which is not selectively sequentially pseudocompact.

As can be seen from Figure 1, the topological groups from Corollary 4 are selectively pseudocompact. Therefore, the following particular version of Corollary 4 (ii) provides a positive answer to Question 1 (i).

Corollary 5. *There exists a selectively pseudocompact Boolean group (of weight $\mathfrak{c}$) which is not selectively sequentially pseudocompact.*

Our next remark clarifies the strength of the condition "without infinite separable pseudocompact subsets" appearing in Theorem 1 and its Corollaries 1 and 3. Indeed, this remark shows that the topological groups in these results contain no infinite subsets which belong to any of the following classes of spaces:

- countably pseudocompact;
- countably pracompact;
- countably compact;
- compact.

Remark 2. *(i) Hernández and Macario [9] say that a space X is* countably pseudocompact *if, for every countable subset A of X, there exists a countable subset B of X such that $A \subseteq \overline{B}$ and $\overline{B}$ is pseudocompact. (Here $\overline{B}$ denotes the closure of B in X.) It is immediately obvious from this definition that every infinite countably pseudocompact space contains an infinite separable pseudocompact subset.*

(ii) A space X is said to be countably pracompact *if X contains a dense set Y such that every infinite subset of Y has an accumulation point in X; see ([10], Ch. III, Sec. 4). Let X be an infinite countably pracompact space, and let Y be its dense subspace such that every infinite subset of Y has an accumulation point in X. Since X is infinite and Y is dense in X, the set Y must be infinite. Fix a countably infinite subset S of Y. Then $C = \overline{S}$ is a separable space. Note that every infinite subset of S has an accumulation point in C. Since S is dense in C, it easily follows that C is pseudocompact. We proved that an infinite countably pracompact space contains an infinite separable pseudocompact subset.*

(iii) Since countably compact spaces are countably pracompact, it follows from (ii) that every infinite countably compact space contains an infinite separable pseudocompact subset.

(iv) Since compact spaces are countably compact, it follows from (iii) that every infinite compact space contains an infinite separable pseudocompact subset.

Remark 3. *The topological groups from Theorem 1 and all its corollaries above are (Pontryagin) reflexive. Indeed, the topological group G from Theorem 1 has no infinite separable pseudocompact subsets, so all compact subsets of G are finite by Remark 2 (iv). Since G is pseudocompact, it is reflexive by ([11], Theorem 2.8) (this also* *follows from ([12], Lemma 2.3 and Theorem 6.1)). Finally, the topological groups from all corollaries of Theorem 1 are obtained by application of this theorem, so they inherit their reflexivity from it.*

Corollaries 1 and 3 and Figure 1 suggest the following natural question:

Question 2. *Does there exist an infinite abelian (or even Boolean) strongly $(\beta\mathbb{N} \setminus \mathbb{N})$-bounded group G satisfying one of the following conditions:*

(i) *G is not selectively sequentially pseudocompact;*

(ii) *G does not have non-trivial convergent sequences?*

Since infinite selectively sequentially pseudocompact spaces contain non-trivial convergent sequences by ([5], Proposition 3.1), Item (ii) of this question is stronger than Item (i). By Remark 1 (ii), Item (ii) of the question cannot be further strengthened by requiring all compact subsets of G to be finite.

According to the double arrow in Figure 1, a positive answer to Question 2 (i) would provide an example of a strongly $(\beta\mathbb{N} \setminus \mathbb{N})$-bounded (abelian) group which is not ω-bounded. However, a topological group with these properties can be easily constructed.

Example 1. *Strongly $(\beta\mathbb{N} \setminus \mathbb{N})$-bounded abelian groups need not be ω-bounded. Indeed, let κ be an uncountable cardinal, and let H be a countably infinite subgroup of the torus group $\mathbb{T}$. For every $h \in H$, let $c_h \in \mathbb{T}^\kappa$ be the constant function from κ to $\mathbb{T}$ defined by $c_h(\alpha) = h$ for all $\alpha \in \kappa$. Define $C = \{c_h : h \in H\}$. Let $D = \{f \in \mathbb{T}^\kappa : |\{\alpha \in \kappa : f(\alpha) \neq 0\}| \leq \omega\}$ be the Σ-product in $\mathbb{T}^\kappa$, and let G be the smallest subgroup of $\mathbb{T}^\kappa$ containing $C \cup D$. Note that C is a closed subgroup of G which is not compact. Indeed, if C were compact, its projection H on a fixed coordinate would be compact as well, and as H would be an infinite compact subgroup of $\mathbb{T}$, we would find that $H = \mathbb{T}$, in contradiction to our assumption that H is countable. Since C is a countably infinite non-compact closed subgroup of G, this shows that G is not ω-bounded. Since G has a dense ω-bounded subgroup D, it is strongly $(\beta\mathbb{N} \setminus \mathbb{N})$-bounded.*

Example 1 shows that Arrow 1 in Figure 1 is not reversible, even for topological groups.

The next remark shows that the assumption in Theorem 1 that the cardinal κ satisfies $\kappa^\omega = \kappa$ is essential and cannot be omitted, at least in ZFC.

Remark 4. *Under the Generalized Continuum Hypothesis (GCH), if $\kappa^\omega > \kappa$, then every dense pseudocompact subgroup G of $\mathbb{Z}_2^\kappa$ contains a non-trivial convergent sequence [13]. Further results in this direction can be found in [14].*

4. Free Boolean Groups $B(X)$ and Free Precompact Boolean Groups $FPB(X)$

Let X be a set. The set $B(X) = [X]^{<\omega}$ of all finite subsets of X becomes an abelian group with the symmetric difference $E + F = (E \setminus F) \cup (F \setminus E)$ as its group operation $+$ and the empty set as its zero element. Clearly, if $E, F \in B(X)$ are disjoint, then $E + F = E \cup F$. Each element E of $B(X)$ has order 2, as $E + E = 0$, so $B(X)$ is a Boolean group.

If one abuses notation by identifying an element $x \in X$ with the singleton $\{x\} \in B(X)$, then each element $E \in B(X)$ of the group $B(X)$ admits a unique decomposition $E = \sum_{x \in E} x$, so the set X can be naturally considered as the set of generators of $B(X)$. (Here we agree that $\sum_{x \in \emptyset} x = 0$.)

Every map $f : X \to \mathbb{Z}_2$ has a unique extension $\tilde{f} : B(X) \to \mathbb{Z}_2$ to a homomorphism of $B(X)$ to $\mathbb{Z}_2$ defined by

$$\tilde{f}(E) = \sum_{x \in E} f(x) \text{ for } E \in B(X), \tag{2}$$

where the sum is taken in the group $\mathbb{Z}_2$. Since the variety $\mathcal{A}_2$ of all Boolean groups is generated by the single group $\mathbb{Z}_2$, the group $B(X)$ coincides with the free group in the variety $\mathcal{A}_2$ over a set X [15].

Thus, $B(X)$ is the *free Boolean group* over X. (Note that the trivial group is the free Boolean group over the empty set.)

Recall that a topological group is *precompact* if it is a subgroup of some compact group, or equivalently, if its completion is compact. The class of all precompact Boolean groups forms a variety $\mathscr{V}$ of topological groups [16,17]. Therefore, given a topological space X, there exists the free object $FPB(X)$ of X in $\mathscr{V}$ [18,19] which we shall call the free precompact Boolean group of X.

Definition 5. *Let $\mathscr{V}$ be the variety of all precompact Boolean groups. For a topological space X, a topological group $FPB(X)$ is said to be the* free precompact Boolean group *of X provided it satisfies two properties:*

(i) $FPB(X) \in \mathscr{V}$,
(ii) *there exists a continuous map $\eta_X : X \to FPB(X)$ such that*

    (a)   *$FPB(X)$ is algebraically generated by $\eta_X(X)$, and*
    (b)   *for every continuous map $\varphi : X \to G$ with $G \in \mathscr{V}$, there exists a continuous homomorphism $\hat{\varphi} : FPB(X) \to G$ such that $\varphi = \hat{\varphi} \circ \eta_X$.*

A description of $FPB(X)$ as the reflection of the free (abelian) topological group of a space X in the class $\mathscr{V}$ of precompact Boolean groups can be found in ([20], Section 9). Another description for zero-dimensional spaces X is given in Lemma 2 below. The reason why zero-dimensionality of a space X plays such an important role can be seen from the following lemma.

Lemma 1. *For a topological space X, the following conditions are equivalent:*

(i) *The map η_X from Item (ii) of Definition 5 is a homeomorphic embedding;*
(ii) *X is zero-dimensional.*

Proof. (i) $\to$ (ii) Since $FPB(X) \in \mathscr{V}$ by Definition 5 (i), $FPB(X)$ is a precompact Boolean group. Then the completion K of $FPB(X)$ is a compact Boolean group, so K is zero-dimensional. Since $\eta_X(X) \subseteq FPB(X) \subseteq K$, the subspace $\eta_X(X)$ of K is zero-dimensional as well. Finally, since $\eta_X(X)$ is homeomorphic to X by our assumption, it follows that X is zero-dimensional.

(ii) $\to$ (i) Since X is zero-dimensional, there exists a homeomorphic embedding $\varphi : X \to G$, where $G = \mathbb{Z}_2^\kappa$ for a suitable cardinal κ. Let η_X and $\hat{\varphi}$ be as in Definition 5 (ii). Since $\varphi = \hat{\varphi} \circ \eta_X$ is an injection, so is η_X. Therefore, $\eta_X : X \to \eta_X(X)$ is a bijection, so it has its inverse map $\eta_X^{-1} : \eta_X(X) \to X$. Similarly, since $\varphi : X \to \varphi(X)$ is a bijection, it has its inverse $\varphi^{-1} : \varphi(X) \to X$. Now, $\varphi \circ \eta_X^{-1} = \hat{\varphi} \circ \eta_X \circ \eta_X^{-1} = \hat{\varphi} \restriction_{\eta_X(X)}$ by Definition 5 (ii) (b), so $\eta_X^{-1} = \varphi^{-1} \circ \hat{\varphi} \restriction_{\eta_X(X)}$. Since φ is a homeomorphic embedding, its inverse φ^{-1} is continuous. Since $\hat{\varphi}$ is continuous as well, so is the composition $\eta_X^{-1} = \varphi^{-1} \circ \hat{\varphi} \restriction_{\eta_X(X)}$. Since η_X is continuous by Definition 5 (ii), we conclude that $\eta_X : X \to \eta_X(X)$ is a homeomorphism. We have proved that $\eta_X : X \to FPB(X)$ is a homeomorphic embedding. $\square$

Lemma 2. *Let X be a zero-dimensional topological space and let $\mathscr{F}_X$ be the family of all continuous maps $f : X \to \mathbb{Z}_2$ from X to the group $\mathbb{Z}_2$ endowed with the discrete topology. Consider the initial topology $\mathscr{T}_X$ on $B(X)$ with respect to the family $\tilde{\mathscr{F}}_X = \{\tilde{f} : f \in \mathscr{F}_X\}$ of homomorphisms; that is, the family $\{\tilde{f}^{-1}(z) : f \in \mathscr{F}_X, z \in \mathbb{Z}_2\}$ forms a subase for the topology $\mathscr{T}_X$. Then the topological group $(B(X), \mathscr{T}_X)$ coincides with the free precompact Boolean group $FPB(X)$ of X, as witnessed by the natural inclusion map of X into $B(X)$ (sending each $x \in X$ to $\{x\} \in B(X)$) taken as η_X. Furthermore, $\mathscr{T}_X$ induces on X the original topology of X.*

Proof. First, we check Items (i) and (ii) of Definition 5.

(i) Since $\mathscr{T}_X$ is the initial topology with respect to the family $\tilde{\mathscr{F}}_X$ consisting of homomorphisms into the compact group $\mathbb{Z}_2$, it is precompact. Since $B(X)$ is a Boolean group, we have $(B(X), \mathscr{T}_X) \in \mathscr{V}$.

(ii) Item (a) is clear, as $\eta_X(X) = X$ algebraically generates $B(X)$. To check Item (b), suppose that $G \in \mathscr{V}$ and $\varphi : X \to G$ is a continuous map. It follows from $G \in \mathscr{V}$ that G is a precompact Boolean group, so its completion K is a compact Boolean group. The standard facts of the duality theory imply that K is topologically isomorphic to the Cartesian product $\mathbb{Z}_2^\tau$ for some cardinal τ. Therefore, we can identify G with a subgroup of $\mathbb{Z}_2^\tau$.

Let $\alpha < \tau$ be arbitrary. Consider the projection $\pi_\alpha : \mathbb{Z}_2^\tau \to \mathbb{Z}_2$ on the αth coordinate. Then the composition map $\varphi_\alpha = \pi_\alpha \circ \varphi : X \to \mathbb{Z}_2$ is continuous, so $\varphi_\alpha \in \mathscr{F}_X$. Now $\bar{\varphi}_\alpha \in \mathscr{\tilde{F}}_X$ by our definition of $\mathscr{\tilde{F}}_X$. Since the topology $\mathscr{T}_X$ has the family $\mathscr{\tilde{F}}_X$ as its subbase, it follows that the homomorphism $\bar{\varphi}_\alpha : (B(X), \mathscr{T}_X) \to \mathbb{Z}_2$ is continuous.

Let $\hat{\varphi} : B(X) \to \mathbb{Z}_2^\tau$ be the continuous homomorphism defined by $\hat{\varphi}(E) = (\bar{\varphi}_\alpha(E))_{\alpha<\tau}$ for $E \in B(X)$. Note that $\hat{\varphi}(x) = (\bar{\varphi}_\alpha(x))_{\alpha<\tau} = (\varphi_\alpha(x))_{\alpha<\tau} = (\pi_\alpha(\varphi(x)))_{\alpha<\tau} = \varphi(x)$ for $x \in X$, as each $\bar{\varphi}_\alpha$ extends φ_α. This shows that $\hat{\varphi} \restriction_X = \varphi$. Since $\varphi : X \to G$ is a homomorphism, X algebraically generates $B(X)$, and G is a subgroup of $\mathbb{Z}_2^\tau$, it follows that $\hat{\varphi}(B(X)) \subseteq G$. We have defined a continuous homomorphism $\hat{\varphi} : (B(X), \mathscr{T}_X) \to G$. Since $\eta_X : X \to B(X)$ is the natural inclusion map, from $\hat{\varphi} \restriction_X = \varphi$ we conclude that $\hat{\varphi} \circ \eta_X = \varphi$.

It follows from (i) and (ii) that $(B(X), \mathscr{T}_X)$ coincides with the free precompact Boolean group $FPB(X)$ of X, as witnessed by the natural inclusion map of X into $B(X)$ taken as η_X. Since X is zero-dimensional, from Lemma 1 we conclude that η_X is a homeomorphic embedding, which implies that $\mathscr{T}_X$ induces the original topology on X. $\square$

Definition 6. *We shall say that a subspace Y of a topological space X is $\mathbb{Z}_2$-embedded in X provided that every continuous map $g : Y \to \mathbb{Z}_2$ can be extended to a continuous map $f : X \to \mathbb{Z}_2$.*

Remark 5. *A clopen subset of a topological space is $\mathbb{Z}_2$-embedded in it.*

We finish this section with the lemma which will be needed in the future proofs.

Lemma 3. *Let X be a zero-dimensional space.*

(i) *If Y is a zero-dimensional space and $\varphi : Y \to X$ is a continuous injection, then the continuous homomorphism $\hat{\varphi} : FPB(Y) \to FPB(X)$ extending φ is an injection as well.*
(ii) *If a closed subset Y of X is $\mathbb{Z}_2$-embedded in X, then $FPB(Y)$ is a closed subgroup of $FPB(X)$.*

Proof. (i) It follows from Lemma 2 that, algebraically, $\hat{\varphi} : B(Y) \to B(X)$ and $\hat{\varphi} \restriction_Y = \varphi$. Since φ is an injection, so is $\hat{\varphi}$.

(ii) By Lemma 2, we can identify $FPB(X)$ and $FPB(Y)$ with $(B(X), \mathscr{T}_X)$ and $(B(Y), \mathscr{T}_Y)$, respectively. Since $B(Y) \subseteq B(X)$, it suffices to show that

(a) $\mathscr{T}_X$ induces the topology $\mathscr{T}_Y$ on $B(Y)$, and
(b) $B(Y)$ is $\mathscr{T}_X$-closed in $B(X)$.

In the proof below, we freely use notations from Lemma 2.

(a) Since Y is a subspace of X, one has $\{f \restriction_Y : f \in \mathscr{F}_X\} \subseteq \mathscr{F}_Y$. Since Y is $\mathbb{Z}_2$-embedded in X, from Definition 6 we obtain the inverse inclusion $\mathscr{F}_Y \subseteq \{f \restriction_Y : f \in \mathscr{F}_X\}$. This establishes the equality $\mathscr{F}_Y = \{f \restriction_Y : f \in \mathscr{F}_X\}$, which implies (a) by definition of $\mathscr{T}_X$ and $\mathscr{T}_Y$.

(b) Suppose that $E \in B(X) \setminus B(Y)$. There then exists $x_0 \in E \setminus Y$. Since E is a finite subset of X and Y is $\mathscr{T}_X$-closed in X, the set $F = Y \cup (E \setminus \{x_0\})$ is $\mathscr{T}_X$-closed in X as well. Since X is zero-dimensional, we can find a clopen subset W of X such that $F \subseteq W$ and $x_0 \notin W$. Define the function $f : X \to \mathbb{Z}_2$ by $f(W) \subseteq \{0\}$ and $f(X \setminus W) \subseteq \{1\}$. Since W is clopen in X, we have $f \in \mathscr{F}_X$, which implies $\tilde{f} \in \mathscr{\tilde{F}}_X$.

Therefore, $O = \tilde{f}^{-1}(1) \in \mathscr{T}_X$ by our definition of $\mathscr{T}_X$. Since $E \setminus \{x_0\} \subseteq F \subseteq W \subseteq f^{-1}(0) \subseteq \tilde{f}^{-1}(0)$ and $x_0 \in X \setminus W \subseteq f^{-1}(1) \subseteq \tilde{f}^{-1}(1)$, we have

$$\tilde{f}(E) = \sum_{x \in E} \tilde{f}(x) = \tilde{f}(x_0) + \sum_{x \in E \setminus \{x_0\}} \tilde{f}(x) = 1 + 0 = 1$$

by Equation (2), so $E \in O$. Since $Y \subseteq W \subseteq f^{-1}(0) \subseteq \tilde{f}^{-1}(0)$, Y algebraically generates $B(Y)$ and $\tilde{f}$ is a homomorphism, we obtain $\tilde{f}(B(Y)) \subseteq \{0\}$. This shows that $O \cap B(Y) = \emptyset$. □

We refer the reader to ([21], Section 2) for properties of free precompact (abelian) groups and [22] for those of free precompact Boolean groups.

5. Coherent Maps

Definition 7. *Given sets $P \subseteq \beta\mathbb{N} \setminus \mathbb{N}$ and K, define $X = P \times K \times (\omega + 1)$ and $X^* = P \times K \times \{\omega\}$.*

Definition 8. *Let X be a set as in Definition 7. We shall say that a map $f : X \to \mathbb{Z}_2$ is coherent provided that*

$$\{n \in \omega : f(p,k,n) = f(p,k,\omega)\} \in p \text{ for every } p \in P \text{ and each } k \in K. \tag{3}$$

Note that the map $f : X \to \mathbb{Z}_2$ is coherent if and only if $f(p,k,\omega)$ is a p-limit of the sequence $\{f(p,k,n) : n \in \mathbb{N}\}$ whenever $p \in P$ and $k \in K$.

Definition 9. *We introduce the topology on a set X as in Definition 7 by declaring each point of $X \setminus X^*$ to be isolated and a basic open neighbourhood of a point $(p,k,\omega) \in X^*$ to be of the form $\{(p,k,\omega)\} \cup \{(p,k,n) : n \in F\}$ for a given element $F \in p$.*

Remark 6. *Let X be a topological space from Definition 9.*

(i) *Note that $X_{p,k} = \{(p,k,n) : n \in \omega + 1\}$ for $(p,k) \in P \times K$ is a clopen subset of X, so $X = \bigoplus_{(p,k) \in P \times K} X_{p,k}$ is a topological sum of $X_{p,k}$.*

(ii) *Since each $X_{p,k}$ for $(p,k) \in P \times K$ is a space with a single non-isolated point, it is zero-dimensional. It follows from this and (i) that X is zero-dimensional as well.*

The straightforward verification of the following lemma is left to the reader.

Lemma 4. *Let X be a set as in Definition 7. Then a map $f : X \to \mathbb{Z}_2$ is coherent in the sense of Definition 8 if and only if it is continuous with respect to the topology on X described in Definition 9 and the discrete topology on $\mathbb{Z}_2$.*

We finish this section with two technical lemmas which will be needed in future proofs. The reader can safely skip them during the first pass.

Lemma 5. *Let X and X^* be sets as in Definition 7. Then every map $g : X \setminus X^* \to \mathbb{Z}_2$ admits a unique coherent extension $f : X \to \mathbb{Z}_2$ over X.*

Proof. For fixed $p \in P$ and $k \in K$, we have

$$\{n \in \omega : g(p,k,n) = 0\} \cup \{n \in \omega : g(p,k,n) = 1\} = \omega \in p.$$

Since p is an ultrafilter on ω, there exists a unique $i_{p,k} = 0,1$ such that

$$\{n \in \omega : g(p,k,n) = i_{p,k}\} \in p. \tag{4}$$

Define $f(p,k,\omega) = i_{p,k}$ for every $p \in P$ and $k \in K$. Finally, let $f(p,k,n) = g(p,k,n)$ for all $(p,k,n) \in X \setminus X^* = P \times K \times \omega$. It follows from this definition and Equation (4) that Equation (3) holds; that is, f is coherent by Definition 8. $\square$

Lemma 6. *Let X be a set as in Definition 7. If $P' \subseteq P$, $K' \subseteq K$ and $h \in B(X) \setminus B(P' \times K' \times (\omega + 1))$, then there exists a coherent map $f : X \to \mathbb{Z}_2$ such that $\tilde{f}(B(P' \times K' \times (\omega + 1))) \subseteq \{0\}$ and $\tilde{f}(h) = 1$.*

Proof. Fix a finite set $F \subseteq X$ such that $h = \sum_{(p,k,n) \in F} \{(p,k,n)\}$. It follows from $h \in B(X) \setminus B(P' \times K' \times (\omega + 1))$ that $F \not\subseteq P' \times K' \times (\omega + 1)$, so we can fix

$$(p_0, k_0, n_0) \in F \setminus (P' \times K' \times (\omega + 1)). \tag{5}$$

Since F is finite, there exists $m \in \omega$ such that $(p_0, k_0, n) \notin F$ for all $n \in \omega$ with $n \geq m$. Define $f : X \to \mathbb{Z}_2$ by

$$f(p,k,n) = \begin{cases} 1 & \text{if } p = p_0, k = k_0 \text{ and either } n = n_0 \text{ or } n \geq m \\ 0 & \text{otherwise} \end{cases} \quad \text{for } (p,k,n) \in X. \tag{6}$$

Let $p \in P$ and $k \in K$ be arbitrary. If either $p \neq p_0$ or $k \neq k_0$, then $f(p,k,n) = 0$ for every $n \in \omega + 1$ by Equation (6), so $\omega = \{n \in \omega : f(p,k,n) = f(p,k,\omega) = 0\} \in p$. Suppose now that $p = p_0$ and $k = k_0$. Then

$$\{n \in \omega : n \geq m\} \subseteq \{n \in \omega : f(p_0, k_0, n) = f(p_0, k_0, \omega) = 1\} = N$$

by Equation (6). Since p is a free ultrafilter on ω, we have $\{n \in \omega : n \geq m\} \in p$. This implies that $N \in p$, and therefore, f is coherent by Definition 8.

If $(p,k,n) \in P' \times K' \times (\omega + 1)$, then either $p \neq p_0$ or $k \neq k_0$ by Equation (5), so $f(p,k,n) = 0$ by Equation (6). Therefore, $f(P' \times K' \times (\omega + 1)) \subseteq \{0\}$. Since $\tilde{f}$ is a homomorphism extending f, it easily follows that $\tilde{f}(B(P' \times K' \times (\omega + 1))) \subseteq \{0\}$.

From the choice of m and Equation (6), we conclude that $f(p,k,n) = 0$ for all $(p,k,n) \in F \setminus \{(p_0, k_0, n_0)\}$. Furthermore, $f(p_0, k_0, n_0) = 1$ by Equation (6).

Since $\tilde{f}$ is a homomorphism extending f, we obtain

$$\tilde{f}(h) = \tilde{f}\left(\sum_{(p,k,n) \in F} \{(p,k,n)\} \right) = \sum_{(p,k,n) \in F} \tilde{f}\{(p,k,n)\} = \sum_{(p,k,n) \in F} f(p,k,n) = f(p_0, k_0, n_0) = 1.$$

This finishes the proof of our lemma. $\square$

6. Coherent Splitting Maps and Their Continuity

Definition 10. *Let X be a set. We shall say that a map $f : X \to \mathbb{Z}_2$ splits a subset A of $B(X)$ provided that the set $\{a \in A : \tilde{f}(a) = i\}$ is infinite for each $i \in \mathbb{Z}_2$, where $\tilde{f} : B(X) \to \mathbb{Z}_2$ is the homomorphism defined in Equation (2).*

Clearly, a subset split by some map must be infinite. The converse also holds:

Lemma 7. *For an arbitrary set X, every infinite subset of $B(X)$ can be split by some map $f : X \to \mathbb{Z}_2$.*

This lemma is part of folklore and can be proved by a straightforward induction. It can also be derived from ([23], Lemma 4.1).

The secondary goal of this paper is to prove the following theorem strengthening Lemma 7 by additionally requiring the splitting map to be coherent.

Theorem 2. *If X is a set as in Definition 7, then every infinite subset of $B(X)$ can be split by some coherent map $f : X \to \mathbb{Z}_2$.*

This theorem constitutes the main technical tool in the proof of Theorem 1 in Section 10. The proof of Theorem 2 is postponed until Section 9.

The next corollary provides a topological reformulation of Theorem 2.

Corollary 6. *Let X be a set as in Definition 7 equipped with the topology described in Definition 9. Then for every infinite subset A of the free precompact Boolean group $FPB(X)$ of X, there exists a continuous homomorphism $\pi : FPB(X) \to \mathbb{Z}_2$ such that the set $\{a \in A : \pi(a) = i\}$ is infinite for each $i \in \mathbb{Z}_2$.*

Proof. In this proof, we use notations from Lemma 2. The space X is zero-dimensional by Remark 6 (ii). By Lemma 2, we can identify $FPB(X)$ with $(B(X), \mathscr{T}_X)$. After this identification, we can think of A as being an infinite subset of $B(X)$. By Theorem 2, A is split by some coherent map $f : X \to \mathbb{Z}_2$. By Lemma 4, f is continuous, and so $f \in \mathscr{F}_X$, which implies that $\pi = \hat{f} \in \mathscr{F}_X$. Since $\mathscr{T}_X$ is the initial topology with respect to the family $\mathscr{F}_X$, the map π is $\mathscr{T}_X$-continuous. Recalling our identification of $FPB(X)$ with $(B(X), \mathscr{T}_X)$, we conclude that the homomorphism $\pi : FPB(X) \to \mathbb{Z}_2$ is continuous. Since A is split by f, it follows from this and Definition 10 that π satisfies the conclusion of our corollary. $\square$

Definition 11. *For simplicity, we shall say that a topological space is* elementary *if it is homeomorphic to a subspace of $\beta\mathbb{N}$ of the form $\mathbb{N} \cup \{p\}$, where $p \in \beta\mathbb{N} \setminus \mathbb{N}$.*

Corollary 7. *Let K be a non-empty set. For every $k \in K$, let Y_k be either an at most countable discrete space or an elementary space. Let $Y = \bigoplus_{k \in K} Y_k$ be the topological sum of the family $\{Y_k : k \in K\}$. Then for every infinite subset A of $FPB(Y)$, there exists a continuous homomorphism $h : FPB(Y) \to \mathbb{Z}_2$ such that the set $\{a \in A : h(a) = i\}$ is infinite for each $i \in \mathbb{Z}_2$.*

Proof. Let $P = \beta\mathbb{N} \setminus \mathbb{N}$ and let $X = P \times K \times (\omega + 1)$ be the set as in Definition 7. We equip X with the topology described in Definition 9. In this proof, we use notations from Remark 6 (i).

Fix a free ultrafilter q on $\mathbb{N}$. Let $k \in K$. If Y_k is an at most countable discrete space, then we can fix an injection $\varphi_k : Y_k \to X_{q,k}$ which will obviously be continuous. If Y_k is an elementary space, then Definition 11 allows us to identify the space Y_k with the subspace $\mathbb{N} \cup \{p_k\}$ of $\beta\mathbb{N}$, for a suitable $p_k \in \beta\mathbb{N} \setminus \mathbb{N}$. Now we can fix an injection $\varphi_k : Y_k \to X_{p_k,k}$ which sends each point $n \in \mathbb{N}$ to the point $(p_k, k, n) \in X_{p_k,k}$ and the point $p_k \in Y_k$ to $(p_k, k, \omega) \in X_{p_k,k}$. Clearly, φ_k is a homeomorphism between Y_k and $X_{p_k,k}$.

Let $\varphi : Y = \bigoplus_{k \in K} Y_k \to X$ be the map such that $\varphi \restriction_{Y_k} = \varphi_k$ for every $k \in K$. Since each φ_k is an injection, so is φ. Since each φ_k is continuous, it follows from our definition of φ and Remark 6 (i) that φ is continuous as well.

Clearly, Y is zero-dimensional, and X is zero-dimensional by Remark 6 (ii). Since $\varphi : Y \to X$ is a continuous injection, $\hat{\varphi} : FPB(Y) \to FPB(X)$ is a continuous monomorphism by Lemma 3 (i).

Let A be an infinite subset of $FPB(Y)$. Then $B = \hat{\varphi}(A)$ is an infinite subset of $FPB(X)$. By Corollary 6, we can find a continuous homomorphism $\pi : FPB(X) \to \mathbb{Z}_2$ such that the set $\{b \in B : \pi(b) = i\}$ is infinite for each $i \in \mathbb{Z}_2$. Now the composition $h = \pi \circ \hat{\varphi} : FPB(Y) \to \mathbb{Z}_2$ is the desired homomorphism, as $\hat{\varphi} \restriction_A : A \to B$ is a one-to-one map. $\square$

7. Applications to Free Precompact Boolean Groups of Topological Sums of Maximal Spaces

Definition 12. *Recall that a space is* maximal *if it is non-discrete, yet any strictly stronger topology on it is discrete.*

One easily sees that every maximal space X has exactly one non-isolated point p such that the trace of the filter of neighbourhoods of p on the set $D = X \setminus \{p\}$ of isolated points of X is an ultrafilter on D. In particular, X is zero-dimensional.

Clearly, elementary spaces from Definition 11 are precisely the countably infinite maximal spaces.

Lemma 8. *Let X be either a discrete or a maximal topological space, and let Y be an at most countable closed subspace of X. Then*

(i) *Y is either elementary or discrete, and*

(ii) *Y is $\mathbb{Z}_2$-embedded in X.*

Proof. The conclusion of our lemma is trivial when X is a discrete space. Therefore, from now on we shall assume that X is a maximal space. Let p be the non-isolated point of X. We consider two cases.

Case 1. $p \in Y$. If p is a non-isolated point in Y, then every neighbourhood of p intersects the set $Y \setminus \{p\}$. By the maximality of X, we conclude that Y is a neighbourhood of p in X. This means that Y is clopen in X, and therefore $\mathbb{Z}_2$-embedded in X by Remark 5. Applying maximality of X once again, we conclude that Y is an elementary space.

Suppose now that p is an isolated point of Y. Then Y is discrete and there exists an open subset U of X such that $U \cap Y = \{p\}$. If $g : Y \to \mathbb{Z}_2$ is a continuous map, then the map $f : X \to \mathbb{Z}_2$ defined by

$$f(x) = \begin{cases} g(x) & \text{if } x \in Y \setminus \{p\} \\ g(p) & \text{if } x \in U \\ 0 & \text{otherwise} \end{cases}$$

is continuous and extends g. This shows that Y is $\mathbb{Z}_2$-embedded in X.

Case 2. $p \in X \setminus Y$. Since p is the only non-isolated point of X, all points of Y are isolated in X, so Y is discrete and open in X. Since Y is also closed in X, it is clopen in X, and so $\mathbb{Z}_2$-embedded in X by Remark 5. $\square$

Lemma 9. *Let $X = \bigoplus_{j \in J} X_j$ be the topological sum of a family $\{X_j : j \in J\}$, where each space X_j is either discrete or maximal. Let A be an at most countable subset of $FPB(X)$. Then there exist an at most countable set $K \subseteq J$ and an at most countable closed subspace Y_k of X_k for each $k \in K$ such that the topological sum $Y = \bigoplus_{k \in K} Y_k$ satisfies the following conditions:*

(i) *each Y_k is either elementary or discrete;*

(ii) *$FPB(Y)$ is an at most countable closed subgroup of $FPB(X)$;*

(iii) *every continuous homomorphism $h : FPB(Y) \to \mathbb{Z}_2$ can be extended to a continuous homomorphism $\varphi : FPB(X) \to \mathbb{Z}_2$;*

(iv) *$A \subseteq FPB(Y)$.*

Proof. Since X is zero-dimensional, Lemma 2 allows us to identify $FPB(X)$ with $(B(X), \mathcal{T}_X)$, so we can view A as a subset of $B(X)$. Since A is countable, there exists an at most countable set $S \subseteq X$ such that $A \subseteq B(S)$. Since $X = \bigoplus_{j \in J} X_j$, we can find an at most countable set $K \subseteq J$, and for every $k \in K$ we can fix an at most countable subset Y_k of X_k such that $S \subseteq Y$, where $Y = \bigoplus_{k \in K} Y_k$. Without loss of generality, we may assume that each Y_k contains the unique non-isolated point of X_k whenever X_k is a maximal space. This assumption means that Y_k is closed in X_k for each $k \in K$.

(i) By Lemma 8, each space Y_k is either elementary or discrete, and Y_k is $\mathbb{Z}_2$-embedded in X_k.

(ii) Since Y_k is a closed $\mathbb{Z}_2$-embedded subspace of X_k for every $k \in K$, we conclude that Y is a closed $\mathbb{Z}_2$-embedded subspace of X. Therefore, $FPB(Y)$ is a closed subgroup of $FPB(X)$ by Lemma 3 (ii). Since Y is zero-dimensional, Lemma 2 allows us to identify $FPB(Y)$ with $(B(Y), \mathcal{T}_Y)$. Since Y is at most countable, so is $B(Y)$ and thus $FPB(Y)$.

(iii) Let $h : FPB(Y) \to \mathbb{Z}_2$ be an arbitrary continuous homomorphism. Since the topology of $FPB(Y)$ induces the original topology of Y by Lemma 2, the restriction $g = h \upharpoonright_Y: Y \to \mathbb{Z}_2$ of h to Y is continuous. Since Y is $\mathbb{Z}_2$-embedded in X, we can find a continuous map $f : X \to \mathbb{Z}_2$ extending g. Since $FPB(X)$ coincides with $(B(X), \mathscr{T}_X)$ and $FPB(Y)$ coincides with $(B(Y), \mathscr{T}_Y)$, it follows that $\varphi = \tilde{f}$ is a continuous homomorphism from $FPB(X)$ to $\mathbb{Z}_2$ whose restriction to $FPB(Y)$ coincides with $h = \tilde{g}$.

(iv) Since $A \subseteq B(S)$ and $S \subseteq Y$, we have $A \subseteq B(S) \subseteq B(Y)$. Therefore, we can view A as a subset of $FPB(Y)$. $\square$

Theorem 3. *Let* $X = \bigoplus_{j \in J} X_j$ *be the topological sum of a family* $\{X_j : j \in J\}$, *where each space* X_j *is either discrete or maximal. Then for every infinite subset* A *of* $FPB(X)$, *there exists a continuous homomorphism* $\varphi : FPB(X) \to \mathbb{Z}_2$ *such that the set* $\{a \in A : \varphi(a) = i\}$ *is infinite for each* $i \in \mathbb{Z}_2$.

Proof. Without loss of generality, we may assume that A is countably infinite. Applying Lemma 9 to this A, we can obtain a subspace Y of X as in the conclusion of Lemma 9. By Item (i) of this lemma, we can apply Corollary 7 to find a continuous homomorphism $h : FPB(Y) \to \mathbb{Z}_2$ such that the set $\{a \in A : h(a) = i\}$ is infinite for each $i \in \mathbb{Z}_2$. Applying Item (iii) of Lemma 9, we can find a continuous homomorphism $\varphi : FPB(X) \to \mathbb{Z}_2$ extending h. Since $A \subseteq FPB(Y)$ by Item (iv) of Lemma 9, we have $\{a \in A : \varphi(a) = i\} = \{a \in A : h(a) = i\}$ for each $i \in \mathbb{Z}_2$. $\square$

Lemma 10. *Let* X *be a topological space such that the closure of each at most countable subset of* X *is at most countable. Then every separable pseudocompact subspace* K *of* X *is compact and metrizable. Moreover, if* K *is infinite, then* K *contains a non-trivial convergent sequence.*

Proof. Let K be a separable pseudocompact subset of X. Let S be an at most countable dense subset of K. Then its closure C in X is at most countable by the assumption of our lemma. Since S is dense in K, we have $K \subseteq C$. Thus, K is an at most countable pseudocompact space, so it must be compact. An at most countable compact space is metrizable [24], so K is a metrizable compact space. The last sentence of our lemma follows from the fact that every infinite compact metrizable space has a non-trivial convergent sequence. $\square$

Theorem 4. *Let* $X = \bigoplus_{j \in J} X_j$ *be the topological sum of a family* $\{X_j : j \in J\}$, *where each space* X_j *is either discrete or maximal. Let* $G = FPB(X)$ *be the free precompact Boolean group of* X. *Then all separable pseudocompact subsets of* G *are finite.*

Proof. First, we check that each at most countable subset A of $G = FPB(X)$ has at most countable closure in G. If A is finite, then it is closed in G. Suppose now that A is infinite. Applying Lemma 9 to this A, we can obtain a subspace Y of X as in the conclusion of Lemma 9. By Item (ii) of this lemma, $H = FPB(Y)$ is an at most countable closed subgroup of $FPB(X) = G$. Note that $A \subseteq H$ by Item (iv) of Lemma 9. Therefore, the closure of A in G is contained in the (at most countable) set H.

Let A be a countably infinite subset of G. Applying Theorem 3, we can find a continuous homomorphism $\varphi : G \to \mathbb{Z}_2$ such that the set $\{a \in A : \varphi(a) = i\}$ is infinite for each $i \in \mathbb{Z}_2$. Since φ is continuous, $A_i = \{a \in A : \varphi(a) = i\}$ is a closed subset of A for $i \in \mathbb{Z}_2 = \{0, 1\}$. Since $A = A_0 \cup A_1$ is a partition of A into two disjoint infinite closed sets, A cannot be a convergent sequence. We have proved that G does not contain non-trivial convergent sequences. By Lemma 10, all separable pseudocompact subsets of G are finite. $\square$

The group $G = FPB(X)$ in Theorem 4 is precompact, so its completion H is a compact group. Being compact, the group H contains many non-trivial convergent sequences. Since these non-trivial convergent sequences in H might appear already in its subgroup G, this demonstrates that Theorem 4 is not completely trivial.

8. Discussion

The topic of this paper is related to a long-standing open problem of van Douwen about the existence in ZFC alone of a countably compact group without non-trivial convergent sequences. (The existence of such a group in some additional set-theoretic axioms, such as Continuum Hypothesis (CH) or Martin's Axiom (MA), is well-known.) Indeed, it was noted in ([5], Example 5.7) that a solution to this problem would bring a positive solution to Question 1 (ii) and thus to the weaker Question 1 (i).

The question of the existence of pseudocompact groups without infinite compact subsets (and its weaker version which only prohibits non-trivial convergent sequences) has been studied extensively [12–14,25,26]. For example, Galindo and Macario proved that, under a mild additional set-theoretic assumption beyond ZFC, every pseudocompact abelian group admits a pseudocompact group topology without infinite compact subsets [12]. Corollary 1 contributes to this topic by constructing an abelian topological group without infinite compact subsets (in fact, even without infinite separable pseudocompact subsets) which has a much stronger property than mere pseudocompactness.

Topological groups without infinite compact subsets play a prominent role in Pontryagin duality theory [27] due to the fact that pseudocompact abelian groups without infinite compact subsets are (Pontryagin) reflexive ([11], Theorem 2.8) (this also follows from ([12], Lemma 2.3 and Theorem 6.1)). All topological groups we construct in this paper are reflexive by Remark 3.

The strongest precompact group topology on an abelian group is called its *Bohr topology*. It is a classical result of Glicksberg that the Bohr topology on any abelian group does not have infinite compact subsets [28]; see also ([29], Section 6) for an alternative proof. Since the free precompact Boolean group $FPB(X)$ of a topological space X is precompact, its topology $\mathscr{T}_X$ is weaker than the corresponding Bohr topology, so $\mathscr{T}_X$ can have more compact subsets than the Bohr topology (in which all compact subsets are finite). Note that, when X is discrete, then $\mathscr{T}_X$ coincides with the Bohr topology on $FPB(X)$, so it does not have infinite compact subsets by Glicksberg's result. Our Theorem 4 can be viewed as an extension of Glicksberg's theorem over free precompact Boolean groups $FPB(X)$ of spaces X very close to being discrete (indeed, maximal spaces are one step from being discrete by Definition 12).

The idea of splitting of a given infinite subset A of a discrete abelian group G via a homomorphism φ from G to some target topological group H (usually $\mathbb{Z}_2$ or the torus group $\mathbb{T}$) is a classical technique for producing a group topology on G without non-trivial convergent sequences. Such a splitting is always possible, modulo natural algebraic restrictions on H and A; see [23,29,30]. However, if G is equipped with a non-discrete group topology $\mathscr{T}$, finding a $\mathscr{T}$-continuous homomorphism φ which splits A is a much more difficult task, and the authors are not aware of any known results in this direction. Therefore, our Theorem 3 can be viewed as a first, albeit somewhat modest, contribution to what is undoubtedly quite an interesting topic.

9. Proof of Theorem 2

In this section, we fix a non-empty set $P \subseteq \beta\mathbb{N} \setminus \mathbb{N}$, a non-empty set K and consider sets

$$X = P \times K \times (\omega + 1) \text{ and } X^* = P \times K \times \{\omega\}$$

from Definition 7. We also fix an infinite subset A of $B(X)$.

Lemma 11. *If $X^* \cap (\bigcup A)$ is finite, then some coherent map $f : X \to \mathbb{Z}_2$ splits A.*

Proof. Since $J = X^* \cap (\bigcup A)$ is finite and A is infinite, there exists $I \in [J]^{<\omega}$ such that the set

$$A' = \{a \in A : a \cap X^* = I\} \tag{7}$$

is infinite. Then

$$B = \{a \setminus X^* : a \in A'\} = \{a \setminus I : a \in A'\} \tag{8}$$

is an infinite subset of $B(X \setminus X^*)$. By Lemma 7, there exists a map $g : X \setminus X^* \to \mathbb{Z}_2$ which splits B. Let $f : X \to \mathbb{Z}_2$ be the unique coherent map extending g given by Lemma 5. Clearly, $\tilde{f} \restriction_{B(X \setminus X^*)} = \tilde{g}$. Since $B \subseteq B(X \setminus X^*)$ and g splits B, the map f splits B as well. It follows from this, Equation (8), and Definition 10 that

$$\{a \in A' : \tilde{f}(a \setminus I) = i\} \text{ is infinite for every } i \in \mathbb{Z}_2. \tag{9}$$

Define $j = \tilde{f}(I)$. Clearly, $j \in \mathbb{Z}_2$. It follows from Equations (7) and (8) that $a = (a \setminus I) \cup I$ for every $a \in A'$, so $a = (a \setminus I) + I$ holds in $B(X)$; therefore,

$$\tilde{f}(a) = \tilde{f}(a \setminus I) + \tilde{f}(I) = \tilde{f}(a \setminus I) + j \text{ for } a \in A', \tag{10}$$

as $\tilde{f}$ is a homomorphism. Combining Equations (9) and (10), we conclude that $\{a \in A' : \tilde{f}(a) = i\}$ is infinite for every $i \in \mathbb{Z}_2$. Since $A' \subseteq A$, the same conclusion holds when A' is replaced by A. According to Definition 10, this means that f splits A. □

Definition 13. *We denote by $\mathbb{Q}$ the set of all triples $q = \langle P^q, K^q, f^q \rangle$, where $P^q \in [P]^{<\omega}$, $K^q \in [K]^{<\omega}$ and $f^q : P^q \times K^q \times (\omega + 1) \to \mathbb{Z}_2$ is a coherent map. For $q = \langle P^q, K^q, f^q \rangle, r = \langle P^r, K^r, f^r \rangle \in \mathbb{Q}$, we let $q \leq r$ provided that $P^r \subseteq P^q$, $K^r \subseteq K^q$, and f^q extends f^r.*

One easily sees that $(\mathbb{Q}, \leq)$ is a poset. Clearly, $\langle \emptyset, \emptyset, \emptyset \rangle \in \mathbb{Q}$, so $\mathbb{Q} \neq \emptyset$.

Recall that a set $D \subseteq \mathbb{Q}$ is said to be *dense* in $(\mathbb{Q}, \leq)$ provided that for every $r \in \mathbb{Q}$ there exists $q \in D$ such that $q \leq r$.

Lemma 12. *(i) For every $p \in P$, the set $C_p = \{q \in \mathbb{Q} : p \in P^q\}$ is dense in $(\mathbb{Q}, \leq)$.*
(ii) For every $k \in K$, the set $E_k = \{q \in \mathbb{Q} : k \in K^q\}$ is dense in $(\mathbb{Q}, \leq)$.

Proof. (i) Suppose that $r \in \mathbb{Q} \setminus C_p$. Then $p \in P \setminus P^r$. Note that the extension $f^q : P^q \times K^q \times (\omega + 1) \to \mathbb{Z}_2$ of f^r, obtained by letting $f^q(p, k, n) = 0$ for all $k \in K^q = K^r$ and $n \in \omega + 1$, is coherent. Then $q = \langle P^q, K^q, f^q \rangle \in \mathbb{Q}$. Clearly, $q \in C_p$ and $q \leq r$.

(ii) Suppose that $r \in \mathbb{Q} \setminus E_k$. Then $k \in K \setminus K^r$. Define $P^q = P^r$ and $K^q = K^r \cup \{k\}$. Note that the extension $f^q : P^q \times K^q \times (\omega + 1) \to \mathbb{Z}_2$ of f^r, obtained by letting $f^q(p, k, n) = 0$ for all $p \in P^q = P^r$ and $n \in \omega + 1$, is coherent. Then $q = \langle P^q, K^q, f^q \rangle \in \mathbb{Q}$. Clearly, $q \in E_k$ and $q \leq r$. □

Lemma 13. *If $X^* \cap (\bigcup A)$ is infinite, then for every $B \in [A]^{<\omega}$ and each $i \in \mathbb{Z}_2$, the set*

$$D_{B,i} = \{q \in \mathbb{Q} : \exists a \in A \setminus B \ (a \subseteq P^q \times K^q \times (\omega + 1) \text{ and } \tilde{f}^q(a) = i)\} \tag{11}$$

is dense in $(\mathbb{Q}, \leq)$.

Proof. Let $r \in \mathbb{Q}$, $B \in [A]^{<\omega}$, and $i \in \mathbb{Z}_2$ be arbitrary. We need to find $q \in \mathbb{Q}$ and $a \in A \setminus B$ such that $q \leq r$, $a \subseteq P^q \times K^q \times (\omega + 1)$, and $\tilde{f}^q(a) = i$.

Since B is finite, the intersection $X^* \cap (\bigcup B)$ is also finite. Furthermore, since both P^r and K^r are finite sets, so is the set $P^r \times K^r \times \{\omega\}$. Therefore,

$$F = (X^* \cap (\bigcup B)) \cup (P^r \times K^r \times \{\omega\}) \tag{12}$$

is a finite subset of X^*. By our hypothesis, $X^* \cap (\bigcup A)$ is infinite, so there exists $a \in A$ such that $(a \cap X^*) \setminus F \neq \emptyset$. Fix $p_0 \in P$, $k_0 \in K$, and $a \in A$ such that $(p_0, k_0, \omega) \in a \setminus F$. It follows from this and Equation (12) that $a \in A \setminus B$.

Since a is a finite subset of $X = P \times K \times (\omega + 1)$, there exist finite sets $P^q \subseteq P$ and $K^q \subseteq K$ such that $a \subseteq P^q \times K^q \times (\omega + 1)$. By Lemma 12, without loss of generality, we may also assume that $P^r \subseteq P^q$ and $K^r \subseteq K^q$.

Let $a' = a \cap (P^r \times K^r \times (\omega + 1))$. Then $j = \tilde{f}^r(a') \in \mathbb{Z}_2$ is well-defined. There exists a unique $l \in \mathbb{Z}_2$ such that $j + l = i$. Note that $(p_0, k_0) \in (P^q \times K^q) \setminus (P^r \times K^r)$, so we can define a map $f^q : P^q \times K^q \times (\omega + 1) \to \mathbb{Z}_2$ by

$$f^q(p, k, n) = \begin{cases} f^r(p, k, n) & \text{if } (p, k, n) \in P^r \times K^r \times (\omega + 1) \\ l & \text{if } (p, k) = (p_0, k_0) \text{ and either } n = \omega \text{ or } (p, k, n) \notin a \\ 0 & \text{otherwise} \end{cases} \tag{13}$$

for all $(p, k, n) \in P^q \times K^q \times (\omega + 1)$.

Claim 1. $q = \langle P^q, K^q, f^q \rangle \in \mathbb{Q}$ and $q \leq r$.

Proof. Since $P^q \in [P]^{<\omega}$ and $K^q \in [K]^{<\omega}$ by our construction, we only need to check that the map $f^q : P^q \times K^q \times (\omega + 1) \to \mathbb{Z}_2$ is coherent. Let $p \in P^q$ and $k \in K^q$ be arbitrary. If $(p, k) \in P^r \times K^r$, then

$$\{n \in \omega : f^q(p, k, n) = f^q(p, k, \omega)\} = \{n \in \omega : f^r(p, k, n) = f^r(p, k, \omega)\} \in p$$

by Equation (13) and coherency of f^r. Suppose now that $(p, k) \in (P^q \times K^q) \setminus (P^r \times K^r)$. If $(p, k) \neq (p_0, k_0)$, then $f^q(p, k, n) = 0$ for all $n \in \omega + 1$ by Equation (13), so $\{n \in \omega : f^q(p, k, n) = f^q(p, k, \omega) = 0\} = \omega \in p$. Finally, if $(p, k) = (p_0, k_0)$, then the second line of Equation (13) implies that $f^q(p, k, \omega) = l$ and $f^q(p, k, n) = l$ for all but finitely many $n \in \omega$, as the set a is finite. Therefore, $\{n \in \omega : f^q(p, k, n) = f^q(p, k, \omega)\}$ is a cofinite subset of ω, so it belongs to p, as p is a free ultrafilter on ω. This finishes the check of the inclusion $q \in \mathbb{Q}$.

Finally, note that f^q extends f^r by the first line of Equation (13). It follows from this, $P^r \subseteq P^q$, $K^r \subseteq K^q$, and Definition 13 that $q \leq r$. $\square$

Claim 2. $\tilde{f}^q(a \setminus a') = l$.

Proof. Since $a' = a \cap (P^r \times K^r \times (\omega + 1))$, we have $a \setminus a' \subseteq ((P^q \times K^q) \setminus (P^r \times K^r)) \times (\omega + 1)$, so Equation (13) implies that $f^q(p_0, k_0, \omega) = l$ and $f^q(p, k, n) = 0$ for all $(p, k, n) \in a \setminus (a' \cup \{(p_0, k_0, \omega)\})$. Since $\tilde{f}^q$ is a homomorphism and $(p_0, k_0, \omega) \in a \setminus a'$ by our choice, this implies

$$\tilde{f}^q(a \setminus a') = \sum_{(p, k, n) \in a \setminus a'} \tilde{f}^q(\{p, k, n\}) = \sum_{(p, k, n) \in a \setminus a'} f^q(p, k, n) = f^q(p_0, k_0, \omega) = l.$$

This establishes our claim. $\square$

Claim 3. $q \in D_{B, i}$.

Proof. The only condition in Equation (11) that remains to be checked is the equality $\tilde{f}^q(a) = i$. Since $a' \subseteq P^r \times K^r \times (\omega + 1) \subseteq P^q \times K^q \times (\omega + 1)$, we have $\tilde{f}^q(a') = \tilde{f}^r(a') = j$. Note that $a = (a \setminus a') \cup a'$, so $a = (a \setminus a') + a'$. Since $\tilde{f}^q$ is a homomorphism, $\tilde{f}^q(a) = \tilde{f}^q(a \setminus a') + \tilde{f}^q(a') = l + j = i$ by Claim 2. $\square$

Since $r \in \mathbb{Q}$ was chosen arbitrarily, the conclusion of our lemma follows from Claims 1 and 3. $\square$

We shall need the following folklore lemma.

Lemma 14. *If $\mathscr{D}$ is an at most countable family of dense subsets of a non-empty poset $(\mathbb{Q}, \leq)$, then there exists an at most countable subset $\mathbb{F}$ of $\mathbb{Q}$ such that $(\mathbb{F}, \leq)$ is a linearly ordered set and $\mathbb{F} \cap D \neq \varnothing$ for every $D \in \mathscr{D}$.*

Proof. Since the family $\mathscr{D}$ is at most countable, we can fix an enumeration $\mathscr{D} = \{D_n : n \in \mathbb{N} \setminus \{0\}\}$ of elements of $\mathscr{D}$. Since $\mathbb{Q} \neq \varnothing$, there exists $q_0 \in \mathbb{Q}$. By induction on $n \in \mathbb{N} \setminus \{0\}$, we can choose $q_n \in D_n$

such that $q_n \leq q_{n-1}$; this is possible because D_n is dense in $(\mathbb{Q}, \leq)$. Now $\mathbb{F} = \{q_n : n \in \mathbb{N} \setminus \{0\}\}$ is the desired subset of $\mathbb{Q}$. □

Lemma 15. *If P and K are at most countable sets and $X^* \cap (\bigcup A)$ is infinite, then some coherent map $f : X \rightarrow \mathbb{Z}_2$ splits A.*

Proof. By Lemmas 12 and 13, the family

$$\mathscr{D} = \{C_p : p \in P\} \cup \{E_k : k \in K\} \cup \{D_{B,i} : B \in [A]^{<\omega}, i \in \mathbb{Z}_2\}$$

consists of dense subsets of $(\mathbb{Q}, \leq)$. Since P, K, and A are at most countable, so is $\mathscr{D}$. By Lemma 14, there exists a set $\mathbb{F} = \{q_n : n \in \mathbb{N}\} \subseteq \mathbb{Q}$ such that $q_0 \geq q_1 \geq \cdots \geq q_n \geq q_{n+1} \geq \ldots$ and $\mathbb{F} \cap D \neq \emptyset$ for every $D \in \mathscr{D}$.

We claim that $f = \bigcup\{f^{q_n} : n \in \mathbb{N}\}$ is a coherent map from X to $\mathbb{Z}_2$ splitting A. Since $\mathbb{F}$ intersects each C_p and every E_k, the domain of f coincides with $X = P \times K \times (\omega + 1)$. Since each f^{q_n} is coherent and f extends all f^{q_n}, it easily follows that f is coherent as well.

Suppose that f does not split A. Then the set $B = \{a \in A : \tilde{f}(a) = i\}$ must be finite for some $i \in \mathbb{Z}_2$, so $B \in [A]^{<\omega}$ and thus $D_{B,i} \in \mathscr{D}$. Therefore, $q_n \in D_{B,i}$ for some $n \in \mathbb{N}$. Applying Equation (11), we can find $a \in A \setminus B$ such that $a \subseteq P^{q_n} \times K^{q_n} \times (\omega + 1)$ and $f^{q_n}(a) = i$. Since $f^{q_n} \subseteq f$, this implies $\tilde{f}(a) = f^{q_n}(a) = i$. Therefore, $a \in B$ by the definition of the set B, in contradiction with $a \in A \setminus B$. □

Proof of Theorem 2. Let A be an infinite subset of $B(X)$. Choose a countably infinite subset A' of A. Since $A' \subseteq B(X) = [X]^{<\omega}$, there exist at most countable sets $P' \subseteq P$ and $K' \subseteq K$ such that $A \subseteq B(X')$, where $X' = P' \times K' \times (\omega + 1)$. Combining Lemmas 11 and 15, we can find a coherent map $f' : X' \rightarrow \mathbb{Z}_2$ splitting A'. Let $f : X \rightarrow \mathbb{Z}_2$ be the extension of f' over X obtained by letting f take 0 everywhere on $X \setminus X'$. Clearly, f is a coherent map which splits A'. Since $A' \subseteq A$, f splits A as well. □

10. Proof of Theorem 1

The following lemma is part of set-theoretic folklore. We include its proof only for convenience of the reader.

Lemma 16. *Let S and T be sets such that $1 \leq |S| \leq |T|$ and T is infinite. Then there exists an enumeration $S = \{s_t : t \in T\}$ such that $|\{t \in T : s_t = s\}| = |T|$ for every $s \in S$.*

Proof. Since $1 \leq |S| \leq |T|$, we can fix a surjection $f : T \rightarrow S$. Since T is infinite, we have $|T| = |T \times T|$, so we can fix a bijection $\theta : T \rightarrow T \times T$. Let $\pi : T \times T \rightarrow T$ be the projection on the first coordinate. Define $s_t = f \circ \pi \circ \theta(t)$ for every $t \in T$. We claim that $\{s_t : t \in T\}$ is the desired enumeration. Indeed, let $s \in S$ be arbitrary. Since f is a surjection, $s = f(t_0)$ for some $t_0 \in T$. Since $|\{t_0\} \times T| = |T|$ and θ is a bijection, the set $T' = \theta^{-1}(\{t_0\} \times T) \subseteq T$ satisfies $|T'| = |T|$. Finally, for every $t \in T'$, we have $\pi \circ \theta(t) \in \pi(\theta(T')) \in \pi(\theta(\theta^{-1}(\{t_0\} \times T))) = \pi(\{t_0\} \times T) = \{t_0\}$, so $s_t = f \circ \pi \circ \theta(t) = f(t_0) = s$. □

Fix a cardinal κ such that $\kappa^\omega = \kappa$ and a set K such that $|K| = \kappa$. Let $K = K_0 \cup K_1$ be a partition of K into pairwise disjoint sets K_i such that $|K_i| = \kappa$ for $i = 0, 1$.

Let P be a non-empty subset of $\beta\mathbb{N} \setminus \mathbb{N}$ satisfying $|P| \leq \kappa$. Consider the set

$$X = P \times K \times (\omega + 1) \tag{14}$$

as in Definition 7. Note that $|X| = \kappa$ by Equation (14) and our assumption on K, P, and κ.

For a set S, we denote by $[S]^{\leq\omega}$ the family of at most countable subsets of S and by $[S]^\omega$ the family of all countably infinite subsets of X.

Claim 4. *(i) There exists an enumeration* $[B(X)]^\omega = \{A_\beta : \beta \in K_0\}$ *such that* $|\{\beta \in K_0 : A_\beta = A\}| = \kappa$
for every $A \in [B(X)]^\omega$.
(ii) There exists an enumeration $[P]^{\leq \omega} \times [K]^{\leq \omega} \times B(X) = \{(P_\beta, K_\beta, h_\beta) : \beta \in K_1\}$ *such that* $|\{\beta \in K_1 : P_\beta = P', K_\beta = K', h_\beta = h\}| = \kappa$ *whenever* $P' \in [P]^{\leq \omega}, K' \in [K]^{\leq \omega}$ *and* $h \in B(X)$.

Proof. (i) Note that $|B(X)| = |X^{<\omega}| = |X| = \kappa$ and $|[B(X)]^\omega| = \kappa^\omega = \kappa = |K_0|$ by our assumption on κ and K_0, so we can apply Lemma 16 (with $S = [B(X)]^\omega$ and $T = K_0$) to fix the desired enumeration $[B(X)]^\omega = \{A_\beta : \beta \in K_0\}$.
(ii) Since $|P| \leq \kappa$ and $|K| = |B(X)| = \kappa$, we have $|[P]^{\leq \omega} \times [K]^{\leq \omega} \times B(X)| \leq \kappa^\omega = \kappa = |K_1|$, so the existence of the desired enumeration $[P]^{\leq \omega} \times [K]^{\leq \omega} \times B(X) = \{(P_\beta, K_\beta, h_\beta) : \beta \in K_1\}$ follows from Lemma 16 applied with $S = [P]^{\leq \omega} \times [K]^{\leq \omega} \times B(X)$ and $T = K_1$. $\square$

For every $\beta \in K$, we define a coherent map $f_\beta : X \to \mathbb{Z}_2$ differently depending on whether $\beta \in K_0$ or $\beta \in K_1$.

Case 1. $\beta \in K_0$. In this case, we use a Theorem 2 to fix a coherent map $f_\beta : X \to \mathbb{Z}_2$ splitting A_β.

Case 2. $\beta \in K_1$. If $h_\beta \in B(X) \setminus B(P_\beta \times K_\beta \times (\omega + 1))$, then we use Lemma 6 to fix a coherent map $f_\beta : X \to \mathbb{Z}_2$ such that $\tilde{f}_\beta(B(P_\beta \times K_\beta \times (\omega + 1))) \subseteq \{0\}$ and $\tilde{f}_\beta(h_\beta) = 1$; otherwise, we let f_β to be the constant map sending X to $\{0\}$ (this map is clearly coherent).

Claim 5. *There exist an enumeration* $[K]^\omega = \{I_k : k \in K\}$ *and a sequence* $\{y_{k,n} : n \in \omega\} \subseteq \mathbb{Z}_2^{I_k}$ *for every* $k \in K$ *such that whenever* $I \in [K]^\omega$ *and* $\{y_n : n \in \omega\} \subseteq \mathbb{Z}_2^I$, *one can find* $k \in K$ *with* $I_k = I$ *and* $y_{k,n} = y_n$ *for all* $n \in \mathbb{N}$.

Proof. Let $S = \bigcup\{(\mathbb{Z}_2^I)^\omega : I \in [K]^\omega\}$. (We recall that $(\mathbb{Z}_2^I)^\omega$ denotes the set of all functions from ω to $\mathbb{Z}_2^I$; each such function s can be considered as a sequence $\{s(n) : n \in \omega\}$ of points of $\mathbb{Z}_2^I$.)
Since $|(\mathbb{Z}_2^I)^\omega| = \mathfrak{c} \leq \kappa$ for every $I \in [K]^\omega$, we have $|S| \leq \kappa^\omega = \kappa = |K|$ by our assumption on κ. Therefore, we can apply Lemma 16 with $T = K$ to fix an enumeration $S = \{s_k : k \in K\}$ such that $\{k \in K : s_k = s\}$ has cardinality κ for every $s \in S$.
Let $k \in K$. Then $s_k \in S$, so $s_k \in (\mathbb{Z}_2^I)^\omega$ for a unique $I \in [K]^\omega$; that is, s_k is a function from ω to $\mathbb{Z}_2^I$. We define $I_k = I$ and $y_{k,n} = s_k(n)$ for all $n \in \omega$.
Let $I \in [K]^\omega$ and $\{y_n : n \in \omega\} \subseteq \mathbb{Z}_2^I$ be arbitrary. Then the function $s : \omega \to \mathbb{Z}_2^I$, defined by $s(n) = y_n$ for $n \in \omega$, belongs to S. By the choice of our enumeration, the set $\{k \in K : s_k = s\}$ has cardinality κ. In particular, there exists $k \in K$ such that $s = s_k$. Now $I_k = I$ and $y_n = s(n) = s_k(n) = y_{k,n}$ for every $n \in \omega$. $\square$

Define
$$y_{p,k,n} = y_{k,n} \text{ for all } (p,k,n) \in P \times K \times \omega. \tag{15}$$

For each $(p,k) \in P \times K$, the sequence $\{y_{p,k,n} : n \in \omega\} = \{y_{k,n} : n \in \omega\}$ of points of the compact space $\mathbb{Z}_2^{I_k}$ has a p-limit $y_{p,k,\omega} \in \mathbb{Z}_2^{I_k}$.
For each $(p,k,n) \in X$, define $z_{p,k,n} \in \mathbb{Z}_2^K$ by

$$z_{p,k,n}(\beta) = \begin{cases} y_{p,k,n}(\beta) & \text{if } \beta \in I_k \\ f_\beta(p,k,n) & \text{if } \beta \in K \setminus I_k \end{cases} \qquad \text{for every } \beta \in K. \tag{16}$$

Claim 6. *For every* $p \in P$ *and each sequence* $\{W_n : n \in \mathbb{N}\}$ *of non-empty open subsets of* $\mathbb{Z}_2^K$, *there exists* $k \in K$ *such that*

(i) $z_{p,k,n} \in W_n$ *for all* $n \in \mathbb{N}$, *and*
(ii) $z_{p,k,\omega}$ *is a* p-limit of the sequence $\{z_{p,k,n} : n \in \mathbb{N}\}$.

Proof. Fix $p \in P$ and a sequence $\{W_n : n \in \mathbb{N}\}$ of non-empty open subsets of $\mathbb{Z}_2^K$. Without loss of generality, we may assume that each W_n is a basic open subset of $\mathbb{Z}_2^K$; that is, $W_n = \prod_{\beta \in K} W_{\beta,n}$, where each $W_{\beta,n}$ is a non-empty (open) subset of $\mathbb{Z}_2$ and $\mathrm{supp}(W_n) = \{\beta \in K : W_{\beta,n} \neq \mathbb{Z}_2\}$ is a finite subset of K. Then the set $J = \bigcup_{n \in \mathbb{N}} \mathrm{supp}(W_n)$ is at most countable, so we can fix a countably infinite subset I of K containing J. For every $n \in \mathbb{N}$, $V_n = \prod_{\beta \in I} W_{\beta,n}$ is a non-empty subset of $\mathbb{Z}_2^I$, so we can select $y_n \in V_n$. By Equation (15) and Claim 5, there exists $k \in K$ such that $I_k = I$ and $y_{p,k,n} = y_{k,n} = y_n$ for all $n \in \mathbb{N}$.

(i) Fix $n \in \mathbb{N}$. By Equation (16), we have

$$z_{p,k,n}(\beta) = y_{p,k,n}(\beta) = y_n(\beta) \in W_{\beta,n} \text{ for every } \beta \in I_k = I. \tag{17}$$

Since $\mathrm{supp}(W_n) \subseteq I$, this implies $z_{p,k,n} \in W_n$.

(ii) It suffices to check that $z_{p,k,\omega}(\beta)$ is a p-limit of the sequence $\{z_{p,k,n}(\beta) : n \in \mathbb{N}\}$ for every $\beta \in K$. We consider two cases.

Case 1. $\beta \in I_k$. Since the sequence $\{y_{p,k,n} : n \in \mathbb{N}\} \subseteq \mathbb{Z}_2^{I_k}$ has a p-limit $y_{p,k,\omega} \in \mathbb{Z}_2^{I_k}$, it follows that $y_{p,k,\omega}(\beta)$ is a p-limit of the sequence $\{y_{p,k,n}(\beta) : n \in \mathbb{N}\}$. Since $\beta \in I_k$, we have $z_{p,k,\omega}(\beta) = y_{p,k,\omega}(\beta)$ by Equation (16). Combining this with Equation (17), we obtain the desired conclusion.

Case 2. $\beta \in K \setminus I_k$. In this case, it follows from Equation (16) that $z_{p,k,n}(\beta) = f_\beta(p,k,n)$ for every $n \in \omega + 1$, and the conclusion follows from the fact that f_β is coherent. $\square$

Claim 7. *The set*

$$Z = \{z_{p,k,n} : (p,k,n) \in X\} \tag{18}$$

is dense in $\mathbb{Z}_2^K$.

Proof. Consider an arbitrary non-empty open subset U of $\mathbb{Z}_2^K$. Let $W_n = U$ for every $n \in \mathbb{N}$. Since P is non-empty, we can choose $p \in P$. Let $k \in K$ be as in the conclusion of Claim 6 applied to this p and the sequence $\{W_n : n \in \mathbb{N}\}$. Then $z_{p,k,1} \in W_1 = U$. Since $(p,k,1) \in X$ by Equation (14), we obtain $z_{p,k,1} \in Z$ by Equation (18), so $Z \cap U \neq \emptyset$. $\square$

Claim 8. *Z is strongly P-pseudocompact.*

Proof. By Definition 4 (ii), we need to check that Z is strongly p-pseudocompact for every $p \in P$. Fix $p \in P$. Let $\{U_n : n \in \mathbb{N}\}$ be a sequence of non-empty open subsets of Z. Since Z is a subspace of $\mathbb{Z}_2^K$, for every $n \in \mathbb{N}$, there exists an open subset W_n of $\mathbb{Z}_2^K$ such that $U_n = Z \cap W_n$; in particular, W_n is non-empty. Let $k \in K$ be as in the conclusion of Claim 6 applied to p and the sequence $\{W_n : n \in \mathbb{N}\}$. By Item (i) of this claim, we have $z_{p,k,n} \in W_n$ for every $n \in \mathbb{N}$. Since $(p,k,n) \in X$ by Equation (14), $z_{p,k,n} \in Z$ by Equation (18), so $z_{p,k,n} \in Z \cap W_n = U_n$ for every $n \in \mathbb{N}$. By Item (ii) of Claim 6, $z_{p,k,\omega}$ is a p-limit of the sequence $\{z_{p,k,n} : n \in \mathbb{N}\}$. Since $(p,k,\omega) \in X$ by Equation (14), we obtain $z_{p,k,\omega} \in Z$ by Equation (18). According to Definition 1, this shows that Z is strongly p-pseudocompact. $\square$

Let G be the subgroup of $\mathbb{Z}_2^K$ generated by Z. Let $f : X \to Z \subseteq G$ be the map defined by

$$f(p,k,n) = z_{p,k,n} \text{ for every } (p,k,n) \in X. \tag{19}$$

Since G is a Boolean group, there exists a unique homomorphism $\tilde{f} : B(X) \to G$ extending f. Since $f(X) = Z$ and the latter set algebraically generates G, the homomorphism $\tilde{f}$ is surjective.

Claim 9. *For every at most countable set $A \subseteq B(X)$, there exists an at most countable set $I \subseteq K$ such that*

$$\pi_\beta \circ \tilde{f}(a) = \tilde{f}_\beta(a) \text{ whenever } \beta \in K \setminus I \text{ and } a \in A, \tag{20}$$

where $\pi_\beta : \mathbb{Z}_2^K \to \mathbb{Z}_2$ is the projection on β'th coordinate.

Proof. For every $a \in A$, there exists a finite set $E_a \subseteq X$ such that

$$a = \sum_{(p,k,n) \in E_a} \{(p,k,n)\}. \tag{21}$$

Since A is at most countable, so is the set

$$J = \{k \in K : \exists p \in P \exists n \in (\omega+1) \, (p,k,n) \in \bigcup \{E_a : a \in A\}\}. \tag{22}$$

Therefore, $I = \bigcup_{k \in J} I_k$ is an at most countable subset of K.

Let $a \in A$ and $\beta \in K \setminus I$ be arbitrary. Suppose that $(p,k,n) \in E_a$. Then $k \in J$ by Equation (22). Therefore, $I_k \subseteq I$ by our choice of I. Since $\beta \notin I$, we conclude that $\beta \in K \setminus I_k$; thus, $z_{p,k,n}(\beta) = f_\beta(p,k,n)$ by Equation (16). Since this holds for every $(p,k,n) \in E_a$ and $\tilde{f}_\beta$ is a homomorphism, from Equations (19) and (21) we conclude that

$$\tilde{f}_\beta(a) = \tilde{f}_\beta \left(\sum_{(p,k,n) \in E_a} \{(p,k,n)\} \right) = \sum_{(p,k,n) \in E_a} \tilde{f}_\beta(\{(p,k,n)\}) = \sum_{(p,k,n) \in E_a} f_\beta(p,k,n) = \sum_{(p,k,n) \in E_a} z_{p,k,n}(\beta)$$

$$= \sum_{(p,k,n) \in E_a} f(p,k,n)(\beta) = \tilde{f} \left(\sum_{(p,k,n) \in E_a} \{(p,k,n)\} \right) (\beta) = \tilde{f}(a)(\beta) = \pi_\beta \circ \tilde{f}(a).$$

This proves Equation (20). $\square$

Claim 10. *G contains no non-trivial convergent sequences.*

Proof. Consider an arbitrary countably infinite set $S \subseteq G$. Since $\tilde{f} : B(X) \to G$ is a surjection, we can fix a countably infinite set $A \subseteq B(X)$ such that $\tilde{f}(A) = S$ and $\tilde{f} \restriction_A : A \to S$ is a bijection. Let $I \subseteq K$ be the set as in the conclusion of Claim 9 (applied to our A). Since $A \in [B(X)]^\omega$, we can apply Claim 4 (i) to conclude that the set $|\{\beta \in K_0 : A_\beta = A\}|$ has cardinality κ. Since $|K_0| = \kappa \geq \mathfrak{c} > \omega \geq |I|$, there exists $\beta \in K_0 \setminus I$. Then f_β splits the set $A = A_\beta$ by our choice of f_β. This means that the set $A_i = \{a \in A : \tilde{f}_\beta(a) = i\}$ is infinite for both $i \in \mathbb{Z}_2$.

Let $i \in \mathbb{Z}_2$ be arbitrary. Since $\tilde{f} \restriction_A : A \to S$ is a bijection, the set $S_i = \tilde{f}(A_i) \subseteq S$ is infinite. It follows from Equation (20) that $\pi_\beta \circ \tilde{f}(a) = \tilde{f}_\beta(a) = i$ for $a \in A_i$, so $\pi_\beta(s) = i$ for $s \in S_i$. Since the map π_β is continuous, it follows that S_i is a closed subset of S.

Since $\tilde{f} \restriction_A : A \to S$ is a bijection and $A = A_0 \cup A_1$ is a partition of A into disjoint sets A_i, it follows that $S = S_0 \cup S_1$ is a partition of S into disjoint sets S_i. Since each S_i is infinite and closed in S, this implies that S cannot be a convergent sequence in G. $\square$

Claim 11. *If $P' \in [P]^{\leq \omega}$ and $K' \in [K]^{\leq \omega}$, then the subgroup $H_{P',K'}$ of G generated by the set*

$$Z_{P',K'} = \{z_{p,k,n} : p \in P', k \in K', n \in \omega+1\} \tag{23}$$

is closed in G.

Proof. Fix $P' \in [P]^{\leq \omega}$ and $K' \in [K]^{\leq \omega}$. Note that $\tilde{f}(B(P' \times K' \times (\omega+1))) = H_{P',K'}$ by Equations (19) and (23).

Let $g \in G \setminus H_{P',K'}$ be arbitrary. Since $\tilde{f}$ is surjective, $\tilde{f}(h) = g$ for some $h \in B(X)$. Clearly, $h \notin B(P' \times K' \times (\omega+1))$. Apply Claim 9 to at most countable subset

$$A = B(P' \times K' \times (\omega+1)) \cup \{h\} \tag{24}$$

of $B(X)$ to obtain at most countable set $I \subseteq K$ as in the conclusion of this claim. Since $(P', K', h) \in [P]^{\leq \omega} \times [K]^{\leq \omega} \times B(X)$, we can apply Claim 4 (ii) to conclude that the set $K_1' = \{\beta \in K_1 : P_\beta = P', K_\beta = K', h_\beta = h\}$ has cardinality κ. Since $|K_1'| = \kappa \geq \mathfrak{c} > \omega \geq |I|$, there exists $\beta \in K_1' \setminus I$. Then $P_\beta = P', K_\beta = K'$ and $h_\beta = h$. Since $h_\beta = h \in B(X) \setminus B(P' \times K' \times (\omega + 1)) = B(X) \setminus B(P_\beta \times K_\beta \times (\omega + 1))$ by our assumption, it follows from $\beta \in K_1' \subseteq K_1$ and our choice of f_β that $\tilde{f}_\beta(B(P' \times K' \times (\omega + 1))) \subseteq \{0\}$ and $\tilde{f}_\beta(h) = 1$. From this, $\beta \in K \setminus I$, Equations (20) and (24), we conclude that $\pi_\beta \circ \tilde{f}(B(P' \times K' \times (\omega + 1))) \subseteq \{0\}$ and $\pi_\beta \circ \tilde{f}(h) = 1$. Since $H_{P',K'} = \tilde{f}(B(P' \times K' \times (\omega + 1)))$ and $g = \tilde{f}(h)$, we get $\pi_\beta(H_{P',K'}) \subseteq \{0\}$ and $\pi_\beta(g) = 1$. Since π_β is continuous, $U_g = \pi_\beta^{-1}(1)$ is an open neighbourhood of g in G disjoint from $H_{P',K'}$.

For every $g \in G \setminus H_{P',K'}$, we found an open neighbourhood U_g of G such that $U_g \cap H_{P',K'} = \varnothing$. Therefore, $H_{P',K'}$ is closed in G. □

Claim 12. *The closure of each at most countable subset of G is at most countable.*

Proof. Let S be an at most countable subset of G. Since Z algebraically generates G, from Equations (18) and (23) we conclude that there exist $P' \in [P]^{\leq \omega}$ and $K' \in [K]^{\leq \omega}$ such that $S \subseteq H_{P',K'}$. (Recall that $H_{P',K'}$ is algebraically generated by $Z_{P',K'}$.) Since $H_{P',K'}$ is closed in G by Claim 11, the closure of S is contained in $H_{P',K'}$. Since P' and K' are at most countable, so is $Z_{P',K'}$ and thus $H_{P',K'}$ as well. □

Claim 13. *All separable pseudocompact subsets of G are finite.*

Proof. This follows from Claims 10 and 12 and Lemma 10. □

Since $Z \subseteq G \subseteq \mathbb{Z}_2^K$, and Z is dense in $\mathbb{Z}_2^K$ by Claim 7, Z is dense in G. Since Z is strongly P-pseudocompact by Claim 8, so is G. By Claim 13, G does not contain infinite separable pseudocompact subsets. Finally, since $|K| = \kappa$, the topological groups $\mathbb{Z}_2^K$ and $\mathbb{Z}_2^\kappa$ are topologically isomorphic.

11. Further Open Questions

In this section we list natural open questions (besides Question 2) inspired by our results.

As was mentioned in Section 8, Galindo and Macario proved that, under a mild additional set-theoretic assumption beyond ZFC, every pseudocompact abelian group admits a pseudocompact group topology without infinite compact subsets [12]. Question 4 below asks for an analogue of their result for other compactness-like properties listed on the left side of Figure 1, while Question 3 is a version of Question 4 restricted to non-trivial convergent sequences. Item (iv) was excluded in Question 4 due to Remark 1 (ii).

Question 3. *Let $\mathcal{P}$ be one of the following properties:*

(i) *selectively pseudocompact;*
(ii) *strongly p-pseudocompact for some $p \in \beta\mathbb{N} \setminus \mathbb{N}$;*
(iii) *strongly $(\beta\mathbb{N} \setminus \mathbb{N})$-pseudocompact;*
(iv) *strongly $(\beta\mathbb{N} \setminus \mathbb{N})$-bounded.*

If an infinite abelian group admits a group topology with property $\mathcal{P}$, must it also admit a group topology with property $\mathcal{P}$ having no non-trivial convergent sequences?

Question 4. *Let $\mathcal{P}$ be one of the properties (i)–(iii) from Question 3. If an infinite abelian group admits a group topology with property $\mathcal{P}$, must it also admit a group topology with property $\mathcal{P}$ having no infinite compact subsets (or even without infinite separable pseudocompact subsets)?*

It makes no sense to ask Questions 3 and 4 for properties on the right side of Figure 1, because infinite selectively sequentially pseudocompact spaces contain non-trivial convergent sequences ([5], Proposition 3.1).

Question 5. *If an abelian group admits a pseudocompact group topology, must it also admit a group topology having one of the stronger properties (i)–(iv) listed in Question 3?*

The version of Question 5 for "selective pseudocompactness" is due to García-Ferreira and Tomita ([7], Question 2.7).

Our last question is related to the reversibility of Arrow 4 in Figure 1 in the class of topological groups.

Question 6. *Does there exist a ZFC example of a selectively pseudocompact (abelian) group which is not strongly p-pseudocompact for any free ultrafilter p on $\mathbb{N}$?*

An example under CH is mentioned in the text after Figure 1.

Author Contributions: Both authors contributed equally to this research work.

Funding: The first listed author was partially supported by the Grant-in-Aid for Scientific Research (C) No. JP26400091 of the Japan Society for the Promotion of Science (JSPS). The second listed author was partially supported by the 2016/2017 fiscal year grant of the Matsuyama Saibikai.

Conflicts of Interest: The authors declare no conflict of interest.

References

1. Bernstein, A.R. A new kind of compactness for topological spaces. *Fund. Math.* **1970**, *66*, 185–193. [CrossRef]
2. Angoa, J.; Ortiz-Castillo, Y.F.; Tamariz-Mascarúa, A. Compact-like properties in hyperspaces. *Matematički Vesnik* **2013**, *65*, 306–318.
3. Angoa, J.; Ortiz-Castillo, Y.F.; Tamariz-Mascarúa, A. Ultrafilters and properties related to compactness. *Topol. Proc.* **2014**, *43*, 183–200.
4. García-Ferreira, S.; Ortiz-Castillo, Y.F. Strong pseudocompact properties. *Comment. Math. Univ. Carol.* **2014**, *55*, 101–109.
5. Dorantes-Aldama, A.; Shakhmatov, D. Selective sequential pseudocompactness. *Topol. Appl.* **2017**, *222*, 53–69. [CrossRef]
6. García-Ferreira, S.; Tomita, A.H. Finite powers of selectively pseudocompact groups. *Topol. Appl.* **2018**, *248*, 50–58. [CrossRef]
7. García-Ferreira, S.; Tomita, A.H. A pseudocompact group which is not strongly pseudocompact. *Topol. Appl.* **2015**, *192*, 138–144. [CrossRef]
8. Shakhmatov, D.; Yañez, V. The impact of the Bohr topology on selective pseudocompactness. *arXiv* **2018**, arXiv:1801.09380.
9. Hernández, S.; Macario, S. Dual properties in totally bounded Abelian groups. *Arch. Math.* **2003**, *80*, 271–283. [CrossRef]
10. Arhangel'skiĭ, A.V. *Topological Function Spaces*; Kluwer: Dordrecht, The Netherlands, 1991; Volume 78.
11. Ardanza-Trevijano, S.; Chasco, M.J.; Domínguez, X.; Tkachenko, M. Precompact noncompact reflexive Abelian groups. *Forum Math.* **2012**, *24*, 289–302. [CrossRef]
12. Galindo, J.; Macario, S. Pseudocompact group topologies with no infinite compact subsets. *J. Pure Appl. Algebra* **2011**, *215*, 655–663. [CrossRef]
13. Malykhin, V.I.; Shapiro, L.B. Pseudocompact groups without convergent sequences. *Math. Notes Acad. Sci. USSR* **1985**, *37*, 59–62. (In Russian)
14. Dijkstra, J.J.; van Mill, J. Groups without convergent sequences. *Topol. Appl.* **1996**, *74*, 275–282. [CrossRef]
15. Dikranjan, D.; Shakhmatov, D. Algebraic structure of pseudocompact groups. *Mem. Am. Math. Soc.* **1998**, *133*, 633. [CrossRef]
16. Higman, G. Unrestricted free products, and varieties of topological groups. *J. Lond. Math. Soc.* **1952**, *27*, 73–81. [CrossRef]

17. Morris, S.A. Varieties of topological groups. *Bull. Aust. Math. Soc.* **1969**, *1*, 145–160. [CrossRef]
18. Comfort, W.W.; van Mill, J. On the existence of free topological groups. *Topol. Appl.* **1988**, *29*, 245–269. [CrossRef]
19. Morris, S.A. Varieties of topological groups: A survey. *Colloq. Math.* **1982**, *46*, 147–165. [CrossRef]
20. Shakhmatov, D.; Spěvak, J. Group valued continuous functions with the topology of pointwise convergence. *Topol. Appl.* **2010**, *157*, 1518–1540. [CrossRef]
21. Shakhmatov, D. Imbeddings into topological groups preserving dimensions. *Topol. Appl.* **1990**, *36*, 181–204. [CrossRef]
22. Sipacheva, O. Free Boolean topological groups. *Axioms* **2015**, *4*, 492–517. [CrossRef]
23. Tkachenko, M.G.; Yaschenko, I. Independent group topologies on Abelian groups. *Topol. Appl.* **2002**, *122*, 425–451. [CrossRef]
24. Arhangel'skiĭ, A.V. An addition theorem for the weight of sets lying in bicompacts. *Dokl. Akad. Nauk SSSR* **1959**, *126*, 239–241. (In Russian)
25. Galindo, J.; García-Ferreira, S. Compact groups containing dense pseudocompact subgroups without non-trivial convergent sequences. *Topol. Appl.* **2007**, *154*, 476–490. [CrossRef]
26. Sirota, S.M. The product of topological groups, and extremal disconnectedness. *Sb. Math.* **1969**, *8*, 169–180. (In Russian) [CrossRef]
27. Chasco, M.J.; Dikranjan, D.; Martín-Peinador, E. A survey on reflexivity of abelian topological groups. *Topol. Appl.* **2012**, *159*, 2290–2309. [CrossRef]
28. Glicksberg, I. Uniform boundedness for groups. *Can. J. Math.* **1962**, *14*, 269–276. [CrossRef]
29. Dikranjan, D.; Shakhmatov, D. A Kronecker-Weyl theorem for subsets of abelian groups. *Adv. Math.* **2011**, *226*, 4776–4795. [CrossRef]
30. Dikranjan, D.; Shakhmatov, D. Hewitt–Marczewski–Pondiczery type theorem for abelian groups and Markov's potential density. *Proc. Am. Math. Soc.* **2010**, *138*, 2979–2990. [CrossRef]

![axioms](axioms logo)

MDPI

Article

(L)-Semigroup Sums †

John R. Martin

Department of Mathematics and Statistics, University of Saskatchewan, 106 Wiggins Road, 241 McLean Hall, Saskatoon, SK S7N 5E6, Canada; martin@math.usask.ca

† In this note, all spaces are Hausdorff, and the term map or mapping shall always mean continuous function.

Received: 12 November 2018; Accepted: 17 December 2018; Published: 22 December 2018

Abstract: An (L)-semigroup S is a compact n-manifold with connected boundary B together with a monoid structure on S such that B is a subsemigroup of S. The sum $S + T$ of two (L)-semigroups S and T having boundary B is the quotient space obtained from the union of $S \times \{0\}$ and $T \times \{1\}$ by identifying the point $(x, 0)$ in $S \times \{0\}$ with $(x, 1)$ in $T \times \{1\}$ for each x in B. It is shown that no (L)-semigroup sum of dimension less than or equal to five admits an H-space structure, nor does any (L)-semigroup sum obtained from (L)-semigroups having an Abelian boundary. In particular, such sums cannot be a retract of a topological group.

Keywords: topological group; Lie group; compact topological semigroup; H-space; mapping cylinder; fibre bundle

MSC: 22A15; 54H11; 55P45; 55R10

1. Introduction

An H-space is a space X together with a continuous multiplication $m : X \times X \to X$ and an identity element $e \in X$ such that $m(e, x) = m(x, e) = x$ for all $x \in X$. If, in addition, the multiplication is associative, then X is called a topological monoid. A space together with an associative continuous multiplication is called a topological semigroup. A compact n-manifold S with connected boundary B together with a topological monoid structure such that B is a subsemigroup of S is called an (L)-semigroup in [1], p. 117. Such a topological monoid S can be considered as a mapping cylinder $MC(f)$ of a quotient morphism $f : X \to X/N$ of a compact connected Lie group X where N is a normal sphere subgroup of X (see [1–3]).

In [2], p. 315, it was shown that every commutative n-dimensional (L)-semigroup is a retract of a compact connected Lie group, and if $n \leq 4$, then every n-dimensional (L)-semigroup is a retract of a compact connected Lie group. In this note, it is shown that the sum of two commutative (L)-semigroups cannot be a retract of a topological group, nor can the sum of two n-dimensional (L)-semigroups if $n \leq 5$.

2. (L)-Semigroup Splitting

Let $\mathbb{I} = [0, 1]$ denote the unit interval endowed with the operation of multiplication of real numbers. If $f : X \to Y$ is a mapping between compact spaces, then the mapping cylinder $MC(f)$ is the quotient space obtained by taking the disjoint union of $X \times \mathbb{I}$ and Y and identifying each point $(x, 0) \in X \times \mathbb{I}$ with $f(x) \in Y$. There are natural embeddings $i_X : X \to MC(f)$ and $i_Y : Y \to MC(f)$, so X and Y may be regarded as disjoint closed subspaces of $MC(f)$, and it is easy to check that $i_Y(Y)$ is a strong deformation retract of $MC(f)$. In the special case when Y consists of a single point v, the mapping cylinder is called the cone over X, denoted by $\text{cone}(X)$.

Let $\mathbb{S}^n$ denote the unit n-sphere in Euclidean n-space $\mathbb{R}^n$. Then, in the following result of Mostert and Shields [1], $\text{cone}(\mathbb{S}^n)$, $n = 0, 1, 3$, is homeomorphic to the unit one-ball in the real line $\mathbb{R}^1$, the unit disk $\mathbb{E}^2$ in the complex plane $\mathbb{C}$, the unit four-ball $\mathbb{E}^4$ in the quaternions $\mathbb{H}$, respectively, and is considered to be a topological monoid with the inherited multiplicative structure.

Proposition 1 (Mostert and Shields [1]; also see [2,3]). *Let X be a compact connected Lie group with a closed normal subgroup N such that N is isomorphic to $\mathbb{S}^n$, $n = 0, 1, 3$, and let $f : X \to X/N = Y$ be the quotient morphism. Then:*

(1) $S = MC(f)$ *is a compact manifold with boundary $i_X(X)$ with S being a topological monoid such that $H(S) = i_X(X)$ is the group of units of S with identity $i_X(1_X)$ and $M(S) = i_Y(Y)$ is the minimal ideal of S with identity $i_Y(1_Y)$.*
(2) $S = MC(f)$ *is a locally-trivial fibre bundle over the Lie group $Y = X/N$ as base with fibre $F = \text{cone}(N)$, the unit n-ball for $n = 1, 2, 4$.*

A compact topological monoid S of the above type is called an (L)-semigroup in the literature and S is nonorientable if $N = \mathbb{S}^0$ and orientable if $N = \mathbb{S}^n$, $n = 1, 3$. (Theorem C in [1]).

Let S and T be two (L)-semigroups with boundary B, and let $h : B \to B$ be an autohomeomorphism of B. The quotient space obtained by taking the union of $S \times \{0\}$ and $T \times \{1\}$ and identifying the point $(x, 0)$ in $S \times \{0\}$ with $(h(x), 1)$ in $T \times \{1\}$ for each $x \in B$ is a closed (i.e., compact without boundary) connected n-manifold. Any manifold M obtained in this fashion is said to admit an (L)-semigroup splitting. In the case when h is the identity mapping, we call M the sum of S and T and denote it by $S + T$. If $S = T$, then $S + S = 2S$, the double of the manifold S.

A space X is said to be homogeneous if for every $a, b, \in X$, there is an autohomeomorphism h of X such that $h(a) = b$.

Proposition 2. *If M admits an (L)-semigroup splitting, then M admits the structure of a topological monoid iff M is a Lie group.*

Proof. If M is a Lie group, then it is a topological monoid. Thus, suppose M is a topological monoid. A finite-dimensional homogeneous compact connected monoid admits the structure of a topological group [4]. If, in addition, it is locally contractible, then it must be a Lie group since a compact connected group is a Lie group iff it is locally contractible [5]. Since M is a closed connected n-manifold, the result follows. □

Proposition 3. *Let G be a compact connected Lie group. If M admits an (L)-semigroup splitting, then so does $M \times G$. In particular, if M is an (L)-semigroup sum, then so is $M \times G$.*

Proof. Let M, S, T, and $h : B \to B$ be defined as in the definition of an (L)-semigroup splitting. Then, $S \times G$ and $T \times G$ are (L)-semigroups with $B \times G$ as a boundary, and the correspondence $(x, g) \mapsto (h(x), g)$ determines an autohomeomorphism of $B \times G$. It follows that $M \times G$ admits an (L)-semigroup splitting if M does. In the case when $h = 1_B$, the identity mapping on B, we obtain $M \times G = (S \times G) + (T \times G)$. □

Remark 1. *It is well known that the fundamental group of an H-space is Abelian and that a covering space of an H-space admits an H-space structure (cf. p. 78 and p. 157 in [6]). According to a famous theorem of J.F.Adams [7], the only spheres that are H-space are $\mathbb{S}^n$, $n = 0, 1, 3, 7$, and it follows that $\mathbb{RP}^n$, $n = 0, 1, 3, 7$, are the only real projective n-spaces, which admit H-space structures. We also remark that if a product space is homogeneous, then it admits an H-space structure iff each factor does (Corollary 2.5 in [8]).*

Proposition 4. *Let B be a compact connected Abelian Lie group and let S, and T be (L)-semigroups with boundary B. Then, the sum $S + T$ does not admit an H-space structure.*

Proof. Let $\mathbb{T}^n$ denote the n-torus, which is the product of n copies of the circle group $\mathbb{S}^1$. In the case when $B = \mathbb{T}^1 = \mathbb{S}^1$, the normal sphere subgroups are $\mathbb{S}^0$ and $\mathbb{S}^1$. For the two element subgroups $\mathbb{S}^0$ of $\mathbb{S}^1$, the quotient morphism $f : \mathbb{S}^1 \to \mathbb{S}^1 / \mathbb{S}^0$ yields $MC(f) = \mathbb{M}^2$, the classical Möbius band (see Example 2.3(b) in [2]). When the normal subgroup of $\mathbb{S}^1$ is $\mathbb{S}^1$, the quotient morphism $f : \mathbb{S}^1 \to \mathbb{S}^1 / \mathbb{S}^1 = \{1\}$ yields $MC(f) = \mathbb{E}^2$, the unit disk in the complex plane $\mathbb{C}$ (see Example 2.3(a) in [2]). Thus, the only two-dimensional (L)-semigroup splittings are $2\mathbb{E}^2 = \mathbb{S}^2$, $\mathbb{E}^2 + \mathbb{M}^2 = \mathbb{RP}^2$ and $2\mathbb{M}^2 = \mathbb{K}^2$, the Klein bottle. By Remarks 1, $\mathbb{S}^2$ and $\mathbb{RP}^2$ do not admit H-spaces structures, nor does $\mathbb{K}^2$ since its fundamental group $\Pi_1(\mathbb{K}^2)$ is not Abelian (this follows from the fact that the Abelianization of $\Pi_1(\mathbb{K}^2)$ is $\mathbb{Z} \oplus \mathbb{Z}_2$, the direct sum of the integers and a cyclic group of order two (see [6], p. 135), but $\Pi_1(\mathbb{K}^2)$ must contain a copy of $\Pi_1(\mathbb{T}^2) = \mathbb{Z} \oplus \mathbb{Z}$ since the two-torus $\mathbb{T}^2$ is a double covering space of the Klein bottle $\mathbb{K}^2$).

It follows from Proposition 2.3 that $\mathbb{S}^2 \times \mathbb{T}^n$, $\mathbb{RP}^3 \times \mathbb{T}^n$, and $\mathbb{K}^2 \times \mathbb{T}^n$, $n = 1, 2, \cdots$, are (L)-semigroup sums. Since $\mathbb{E}^2 \times \mathbb{T}^n$ and $\mathbb{M}^2 \times \mathbb{T}^n$ are the only $(n+2)$-dimensional (L)-semigroups with boundary $B = \mathbb{T}^{n+1}$ (see Corollaries 7.5.4 and 7.5.5 in [1]), it follows that the (L)-semigroup sum $S + T$ must be one of the manifolds $\mathbb{S}^2$, $\mathbb{RP}^2$, $\mathbb{K}^2$, $\mathbb{S}^2 \times \mathbb{T}^n$, $\mathbb{RP}^2 \times \mathbb{T}^n$ or $\mathbb{K}^2 \times \mathbb{T}^n$ for $n = 1, 2, \cdots$. However none of these manifolds admit on H-space structure since a homogeneous product space admits an H-space structure iff each of its factors does (see Corollary 2.5 in [8]). $\square$

We remark that a retract of a homogeneous H-space admits an H-space structure (cf. Proposition 2.4 in [8]). Consequently, we have the following corollary.

Corollary 1. *Let B be a compact connected Abelian Lie group, and let S and T be (L)-semigroups with boundary B. Then, the sum $S + T$ is not a retract of a topological group.*

Proposition 5. *If M is a manifold that admits an (L)-semigroup splitting and is either two-dimensional or orientable and three-dimensional, then the following statements are equivalent:*

(1) *M is a retract of a topological group.*
(2) *M admits an H-space structure.*
(3) *M is a Lie group.*

Proof. In the two-dimensional case, the collection of (L)-semigroup sums coincides with the collection of spaces that admit (L)-semigroup splittings since the connected sum of two surfaces is independent of the homeomorphism h used to form the connected sum. Thus, the only surfaces that admit (L)-semigroup splittings are $\mathbb{S}^2$, $\mathbb{RP}^2$, and $\mathbb{K}^2$, and the result follows for surfaces.

The remark following the proof of Proposition 4 shows that (1) implies (2), and since the topological group is a retract of itself, (3) implies (1). Thus, it suffices to show that (2) $\Rightarrow$ (3). As was noted in the proof of Proposition 4, the only orientable three-dimensional (L)-semigroup is the solid torus $\mathbb{E}^2 \times \mathbb{S}^1$. It follows that M must be a (p,q)-lens space $L(p,q)$ where the degenerate cases $L(0,1) = \mathbb{S}^2 \times \mathbb{S}^1$ and $L(1,q) = \mathbb{S}^3$ are included (see p. 234 in [9]). It follows from a theorem of William Browder (p. 140 in [10]) that only $L(1,q) = \mathbb{S}^3$ and $L(2,1) = \mathbb{RP}^3 = SO(3)$ admit H-space structures. Since each of these spaces is a Lie group, the result follows. $\square$

Lemma 1. *Let X be a closed n-manifold, which is the total space of a locally-trivial $\mathbb{S}^2$ fibre bundle over a compact Lie group G. Then, X does not admit an H-space structure.*

Proof. Suppose X does admit an H-space structure, and consider the fibre bundle $\mathbb{S}^2 \to X \to G$. This sequence extends to a fibration sequence $\cdots \Omega G \to \mathbb{S}^2 \to X \to G$ (cf. [11], p. 409). Since X is a (compact metric) ANR-space (see [12]), it has the homotopy type of a finite complex ([13], Corollary 44.2), and it follows from a theorem of W.Browder ([14]) that $\Pi_2(X) = 0$, where $\Pi_2(X)$ denotes the second homotopy group of X. Exactness yields a surjection from $\Pi_2(\Omega G)$ onto $\Pi_2(\mathbb{S}^2)$. An element of $\Pi_2(\Omega G)$ mapping to a generator of $\Pi_2(\mathbb{S}^2)$ is represented by a map $\mathbb{S}^2 \to \Omega G$ whose composition with the map $\Omega(G) \to \mathbb{S}^2$ is homotopic to the identity mapping $1_{\mathbb{S}^2}$ on $\mathbb{S}^2$. Consequently, there is a homotopy retraction $r : \Omega G \to \mathbb{S}^2$ (i.e., $r | \mathbb{S}^2$ is homotopic to $1_{\mathbb{S}^2}$). Since a loop space admits an H-space structure, we may assume that ΩG is an H-space with identity e, and we may assume that $e \in \mathbb{S}^2$ (since ΩG is a homogeneous space when viewed as a loop group).

Define a mapping $m : \mathbb{S}^2 \times \mathbb{S}^2 \to \mathbb{S}^2$ by $m(x, y) = r(xy)$ for $x, y \in \mathbb{S}^2$, where xy denotes the product of x and y in the H-space ΩG. The maps $\mathbb{S}^2 \to \mathbb{S}^2$ given by $x \mapsto m(x, e)$ and $x \mapsto m(e, x)$ are homotopic to the identity mapping $1_{\mathbb{S}^2}$, and therefore, e is a homotopy identity of $\mathbb{S}^2$. For CW complexes the existence of a homotopy identity can be used as the definition of an H-space (see [11], p. 291). Consequently, $\mathbb{S}^2$ admits an H-space structure, and this contradiction completes the proof of the lemma. □

Proposition 6. *Let* $S = MC(f)$ *be an (L)-semigroup as defined in Proposition 1 where X is a compact connected Lie group and* $f : X \to X/N = Y$ *is a quotient morphism with N being a closed normal subgroup of X, which is isomorphic to* $\mathbb{S}^1$. *Then, the double 2S does not admit an H-space structure.*

Proof. By Proposition 1 $S = MC(f)$ is a locally-trivial $\mathbb{E}^2$ bundle over the compact connected Lie group Y, and it follows that its double is a locally-trivial $\mathbb{S}^2$ bundle over Y. Consequently, by Lemma 1, $2S$ does not admit an H-space structure. □

Corollary 2. *Let* $f : U(n) \to U(n)/\mathbb{Z}U(n) = PU(n)$ *denote the quotient morphism where U(n) is the unitary group,* $\mathbb{Z}(U(n))$ *is its centre, and PU(n) is the projective unitary group. Then, if* $n > 1$ *and* $S = MC(f)$, *the double 2S does not admit an (H)-space structure.*

Proof. The elements of $U(n)$ are the complex $n \times n$ unitary matrices, and its centre $\mathbb{Z}(U(n))$ is isomorphic to $\mathbb{S}^1$ since its elements are diagonal matrices equal to $e^{i\theta}$ multiplied by the identity matrix. It follows from Proposition 6 that $2S$ does not admit an H-space structure. □

Theorem 1. *No (L)-semigroup sum of dimension* $n \le 5$ *admits an H-space structure.*

Proof. Proposition 4 shows that the result is true for all n-dimensional (L)-semigroup sums of the form $S + L$ where both S and L have a compact connected Abelian Lie group boundary B. Thus, we need only consider admissible n-dimensional non-Abelian boundaries B with $n = 3, 4$. Hence, B must be one of $\mathbb{S}^3$, $\mathbb{S}^1 \times \mathbb{S}^3$, $\mathbb{S}^1 \times SO(3)$ and $U(2)$ (we note that $SO(3)$ does not qualify as an admissible boundary for an (L)-semigroup since it does not contain normal subgroups of the form $\mathbb{S}^n$, $n = 0, 1, 3$).

In [2], it is shown that the (L)-semigroups with boundary $\mathbb{S}^3$ are $\mathbb{E}^4$ and the four-dimensional Möbius manifold $\mathbb{M}^4$ (which is homeomorphic to $\mathbb{RP}^4$ with the interior of a four-dimensional Euclidean ball removed). It follows (see [2]) that the (L)-semigroups with boundaries $\mathbb{S}^3$, $\mathbb{S}^1 \times \mathbb{S}^3$, $\mathbb{S}^1 \times SO(3)$ are $\mathbb{E}^4$, $\mathbb{M}^4$, $\mathbb{E}^2 \times \mathbb{S}^3$, $\mathbb{M}^2 \times \mathbb{S}^3$, $\mathbb{S}^1 \times \mathbb{E}^4$, $\mathbb{S}^1 \times \mathbb{M}^4$, $\mathbb{E}^2 \times SO(3)$, $\mathbb{M}^2 \times SO(3)$, and the corresponding (L)-semigroup sums are $\mathbb{S}^4$, $\mathbb{RP}^4$, $2\mathbb{M}^4$, $\mathbb{S}^2 \times \mathbb{S}^3$, $\mathbb{RP}^4 \times \mathbb{S}^3$, $\mathbb{K}^2 \times \mathbb{S}^3$, $\mathbb{S}^1 \times \mathbb{S}^4$, $\mathbb{S}^1 \times \mathbb{RP}^4$, $\mathbb{S}^1 \times (2\mathbb{M}^4)$, $\mathbb{S}^2 \times SO(3)$, $\mathbb{RP}^2 \times SO(3)$, $\mathbb{K}^2 \times SO(3)$. Since a retract of a homogeneous H-space admits an H-space structure (cf. [8], Prop. 2.4), it follows that no product containing a copy of $\mathbb{S}^2$, $\mathbb{S}^4$, $\mathbb{RP}^2$, $\mathbb{RP}^4$ of $\mathbb{K}^2$ as a factor can admit an H-space structure. This leaves only $2\mathbb{M}^4$ for consideration. However, its fundamental group $\Pi_1(2\mathbb{M}^4)$ is the free product of $\Pi_1(\mathbb{RP}^4) = \mathbb{Z}_2$ with itself, which is non-Abelian, so $2\mathbb{M}^4$ does not admit an H-space structure. Finally, the only five-dimensional (L)-semigroup sum with boundary $U(2)$ is the manifold $2U(2)$ in Corollary 2, which does not admit an H-space structure. □

Corollary 3. *No (L)-semigroup sum of dimension* $n \le 5$ *is a retract of a topological group.*

Proof. It was noted above that every retract of a homogeneous H-space admits an H-space structure. Since a topological group is an H-space, the result follows from Theorem 1. □

In [15], a space homeomorphic to a retract of a topological group is called a GR-space (often referred to as a retral space in the literature). Clearly AR-spaces and topological groups themselves are GR-spaces, and in [2], it was shown that $\mathbb{M}^2$ and $\mathbb{M}^4$ are GR-spaces. Since GR-spaces are preserved by topological products, it follows that products of $\mathbb{E}^2$, $\mathbb{E}^4$, $\mathbb{M}^2$, $\mathbb{M}^4$, and topological groups are GR-spaces. This will include all the (L)-semigroups mentioned in this note excluding (L)-semigroups with boundary $U(n)$, $n \geq 2$. This suggests two questions.

(a)   Is every (L)-semigroup a retract of a topological group?
(b)   Does every (L)-semigroup sum fail to admit an H-space structure?

Funding: This research received no external funding.

Acknowledgments: I am very indebted to Karl H. Hofmann who introduced and explained (L)-semigroups to me and to John Harper and Allen Hatcher for private communications involving H-spaces.

Conflicts of Interest: The author declares no conflict of interest.

References

1.   Mostert, P.S.; Shields, A.L. On the structure of semigroups on a compact manifold with boundary. *Ann. Math.* **1957**, *65*, 117–143. [CrossRef]
2.   Hofmann, K.H.; Martin, J.R. Möbius manifolds, monoids, and retracts of topological groups. *Semigroup Forum* **2015**, *90*, 301–316. [CrossRef]
3.   Hofmann, K.H.; Mostert, P.S. Elements of Compact Semigroups; Merrill Publishing Company: Columbus, OH, USA, 1966.
4.   Hudson, A.; Mostert, P.S. A finite-dimensional clan is a group. *Ann. Math.* **1963**, *78*, 41–46. [CrossRef]
5.   Hofmann, K.H.; Kramer, L. Transitive actions of locally compact groups on locally contractible spaces. *J. Reine Angew. Math.* **2015**, *702*, 227–243; Erratum in **2015**, *702*, 245–246.
6.   Massey, W.S. Algebraic Topology: An Introduction; Springer-Verlag: New York, NY, USA, 1987.
7.   Adams, J.F. On the non-existence of elements of Hopf invariant one. *Ann. Math.* **1960**, *72*, 20–104. [CrossRef]
8.   Hofmann, K.H.; Martin, J.R. Topological Left-loops. *Topol. Proc.* **2012**, *39*, 185–194.
9.   Rolfsen, D. *Knots and Links*; Mathematics Lecture Series 7; Publish or Perish, Inc.: Berkely, CA, USA, 1976.
10.   Browder, W. The cohomology of covering spaces of H-spaces. *Bull. Am. Math. Soc.* **1959**, *65*, 140–141. [CrossRef]
11.   Hatcher, A. *Algebraic Topology*; Cambridge University Press: Cambridge, UK, 2002.
12.   Borsuk, K. Theory of Retracts. In *Monografie Matematyczne*; PWN: Warsaw, Poland, 1967; Volume 44.
13.   Chapman, T.A. *Lectures on Hilbert Cube Manifolds*; American Mathematical Soc.: Providence, RI, USA, 1976.
14.   Browder, W. Torsion in H-spaces. *Ann. Math.* **1961**, *74*, 24–51. [CrossRef]
15.   Hofmann, K.H.; Martin, J.R. Retracts of topological groups and compact monoids. *Topol. Proc.* **2014**, *43*, 57–67.

axioms

MDPI

Article

Varieties of Coarse Spaces

Igor Protasov [ORCID]

Faculty of Computer Science and Cybernetics, Kyiv University, Academic Glushkov pr. 4d, 03680 Kyiv, Ukraine; i.v.protasov@gmail.com or do@unicyb.kiev.ua

Received: 27 March 2018; Accepted: 10 May 2018; Published: 14 May 2018

Abstract: A class $\mathfrak{M}$ of coarse spaces is called a variety if $\mathfrak{M}$ is closed under the formation of subspaces, coarse images, and products. We classify the varieties of coarse spaces and, in particular, show that if a variety $\mathfrak{M}$ contains an unbounded metric space then $\mathfrak{M}$ is the variety of all coarse spaces.

Keywords: coarse structure; coarse space; ballean; varieties of coarse spaces

MSC: 54E35; 08B85

---

1. Introduction

Following [1], we say that a family $\mathcal{E}$ of subsets of $X \times X$ is a *coarse structure* on a set X if:

- Each $\varepsilon \in \mathcal{E}$ contains the diagonal $\triangle_X$, $\triangle_X = \{(x,x) : x \in X\}$;
- If $\varepsilon, \delta \in \mathcal{E}$ then $\varepsilon \circ \delta \in \mathcal{E}$ and $\varepsilon^{-1} \in \mathcal{E}$, where $\varepsilon \circ \delta = \{(x,y) : \exists z((x,z) \in \varepsilon, (z,y) \in \delta)\}$, and $\varepsilon^{-1} = \{(y,x) : (x,y) \in \varepsilon\}$;
- And if $\varepsilon \in \mathcal{E}$ and $\triangle_X \subseteq \varepsilon' \subseteq \varepsilon$ then $\varepsilon' \in \mathcal{E}$.

Each $\varepsilon \in \mathcal{E}$ is called an *entourage* of the diagonal. A subset $\mathcal{E}' \subseteq \mathcal{E}$ is called a *base* for $\mathcal{E}$ if, for every $\varepsilon \in \mathcal{E}$ there exists $\varepsilon' \in \mathcal{E}'$ such that $\varepsilon \subseteq \varepsilon'$.

The pair $(X, \mathcal{E})$ is called a *coarse space*. For $x \in X$ and $\varepsilon \in \mathcal{E}$, we denote $B(x, \varepsilon) = \{y \in X : (x,y) \in \varepsilon\}$ and say that $B(x, \varepsilon)$ is a *ball of radius ε around x*. We note that a coarse space can be considered as an asymptotic counterpart of a uniform topological space and can be defined in terms of balls, see [2,3]. In this case a coarse space is called a *ballean*.

A coarse space $(X, \mathcal{E})$ is called *connected* if, for any $x, y \in X$, there exists $\varepsilon \in \mathcal{E}$ such that $y \in B(x, \varepsilon)$. A subset Y of X is called *bounded* if there exist $x \in X$ and $\varepsilon \in \mathcal{E}$ such that $Y \subseteq B(x, \varepsilon)$. The coarse structure $\mathcal{E} = \{\varepsilon \in X \times X : \triangle_X \subseteq \varepsilon\}$ is the unique coarse structure such that $(X, \mathcal{E})$ is connected and bounded. In what follows, all coarse spaces under consideration are assumed to be **connected**.

Given a coarse space $(X, \mathcal{E})$, each subset $Y \subseteq X$ has the natural coarse structure $\mathcal{E}|_Y = \{\varepsilon \cap (Y \times Y) : \varepsilon \in \mathcal{E}\}$, where $(Y, \mathcal{E}|_Y)$ is called a *subspace* of $(X, \mathcal{E})$. A subset Y of X is called *large* (or *coarsely dense*) if there exists an $\varepsilon \in \mathcal{E}$ such that $X = B(Y, \varepsilon)$ where $B(Y, \varepsilon) = \cup_{y \in Y} B(Y, \varepsilon)$.

Let $(X, \mathcal{E})$, and $(X', \mathcal{E}')$ be coarse spaces. A mapping $f : X \longrightarrow X'$ is called *coarse* (or *bornologous* in the terminology of [1]) if, for every $\varepsilon \in \mathcal{E}$ there exists an $\varepsilon' \in \mathcal{E}'$ such that, for every $x \in X$, we have $f(B(x, \varepsilon)) \subseteq (B(f(x), \varepsilon'))$. If f is surjective and coarse then $(X', \mathcal{E}')$ is called a *coarse image* of $(X, \mathcal{E})$. If f is a bijection, such that f and f^{-1} are coarse mappings, then f is called an *asymorphism*. The coarse spaces $(X, \mathcal{E})$, $(X', \mathcal{E}')$ are called *coarsely equivalent* if there exist large subsets $Y \subseteq X$, and $Y' \subseteq X$ such that $(Y, \mathcal{E}|_Y)$ and $(Y', \mathcal{E}'|_{Y'})$ are asymorphic.

To conclude the coarse vocabulary, we take a family $\{(X_\alpha, \mathcal{E}_\alpha) : \alpha < \kappa\}$ of coarse spaces and define the *product* $P_{\alpha < \kappa}(X_\alpha, \mathcal{E}_\alpha)$ as the set $P_{\alpha < \kappa} X_\alpha$ endowed with the coarse structure with the base set $P_{\alpha < \kappa} \mathcal{E}_\alpha$. If $\varepsilon_\alpha \in \mathcal{E}_\alpha$, for $\alpha < \kappa$ and $x, y \in P_{\alpha < \kappa} X_\alpha$, where $x = (x_\alpha)_{\alpha < \kappa}$, and $y = (y_\alpha)_{\alpha < \kappa}$ then $(x,y) \in (\varepsilon_\alpha)_{\alpha < \kappa}$ if and only if $(x_\alpha, y_\alpha) \in \varepsilon_\alpha$ for every $\alpha < \kappa$.

Let $\mathfrak{M}$ be a class of coarse spaces closed under asymorphisms. We say that $\mathfrak{M}$ is a *variety* if $\mathfrak{M}$ is closed under the formation of subspaces ($\mathbf{S}\mathfrak{M} \subseteq \mathfrak{M}$), coarse images ($\mathbf{Q}\mathfrak{M} \subseteq \mathfrak{M}$), and products ($\mathbf{P}\mathfrak{M} \subseteq \mathfrak{M}$).

For an infinite cardinal κ, we say that a coarse space $(X, \mathcal{E})$ is κ-*bounded* if every subset $Y \subseteq X$, such that $|Y| < \kappa$, is bounded. Additionally, we denote $\mathfrak{M}_\kappa$ as the variety of all κ-bounded coarse spaces. We denote by $\mathfrak{M}_{single}$ and $\mathfrak{M}_{bound}$ the variety of singletons and the variety of all bounded coarse spaces, respectively. Thus we have the chain of varieties:

$$\mathfrak{M}_{single} \subset \mathfrak{M}_{bound} \subset \ldots \subset \mathfrak{M}_\kappa \subset \ldots \subset \mathfrak{M}_\omega.$$

In Section 2, we prove that every variety of coarse spaces lies in this chain and, in Section 3, we discuss some extensions of this result to coarse spaces endowed with additional algebraic structures.

2. Results

We recall that a family $\mathcal{I}$ of subsets of a set X is an *ideal* in the Boolean algebra $\mathcal{P}_X$ of all subsets of X, if $\mathcal{I}$ is closed under finite unions and subsets. Every ideal $\mathcal{I}$ defines a coarse structure with the base $\{\mathcal{E}_A : A \in \mathcal{I}\}$ where $\mathcal{E}_A = (A \times A) \cup \triangle_X$. Therefore, $B(x, \mathcal{E}_A) = A$ if $x \in A$ and $B(x, \mathcal{E}_A) = \{x\}$ if $x \in X \setminus A$. We denote the obtained coarse space by $(X, \mathcal{I})$. For a cardinal κ, $[X]^{<\kappa}$ denotes the ideal $\{Y \subseteq X : |Y| < \kappa\}$. If $(X, \mathcal{E})$ is a coarse space, the family $\mathcal{I}$ of all bounded subsets of X is an ideal. The coarse space $(X, \mathcal{I})$ is called the *companion* of $(X, \mathcal{E})$.

Let $\mathcal{K}$ be a class of coarse spaces. We say that a coarse space $(X, \mathcal{E})$ is *free* with respect to $\mathcal{K}$ if, for every $(X', \mathcal{E}') \in \mathcal{K}$ every mapping $f : (X, \mathcal{E}) \longrightarrow (X', \mathcal{E}')$ is coarse. For example, $(X, [X]^{<\kappa})$ is free with respect to the variety $\mathfrak{M}_\kappa$. Since $(\kappa, [\kappa]^{<\kappa}) \in \mathfrak{M}_\kappa$ but $(\kappa, [\kappa]^{<\kappa}) \notin \mathfrak{M}_{\kappa'}$ for each $\kappa' > \kappa$; the inclusion $\mathfrak{M}_{\kappa'} \subset \mathfrak{M}_\kappa$ is strict.

Lemma 1. *If a coarse space $(X, \mathcal{E})$ is free with respect to a class $\mathcal{K}$ then $(X, \mathcal{E})$ is free with respect to* $\mathbf{S}\mathcal{K}$, $\mathbf{Q}\mathcal{K}$, $\mathbf{P}\mathcal{K}$.

Proof. We verify only the second statement. Let $(X', \mathcal{E}') \in \mathcal{K}$, $(X'', \mathcal{E}'') \in \mathbf{Q}\mathcal{K}$, and $h : (X', \mathcal{E}') \longrightarrow (X'', \mathcal{E}'')$ be a coarse surjective mapping. We take an arbitrary $f : X \longrightarrow X''$ and choose $h' : X \longrightarrow X'$ such that $f = hh'$. Since $(X, \mathcal{E})$ is free with respect to $\mathcal{K}$, $h' : (X, \mathcal{E}) \longrightarrow (X', \mathcal{E}')$ is coarse so f is coarse as the composition of the coarse mappings h, and h'. $\square$

Lemma 2. *Let X be a set and let $\mathcal{K}$ be a class of coarse spaces, $\mathcal{K} \neq \mathfrak{M}_{single}$. Then there exists a coarse structure $\mathcal{E}$ on X such that $(X, \mathcal{E}) \in \mathbf{SP}\mathcal{K}$ and $(X, \mathcal{E})$ is free with respect to $\mathcal{K}$.*

Proof. We take a set S of all pairwise non-asymorphic coarse spaces $(X', \mathcal{E}') \in \mathcal{K}$, such that $|X'| \leq |X|$, and enumerate all possible triplets $\{(X_\alpha, \mathcal{E}_\alpha, f_\alpha) : \alpha < \lambda\}$, such that $(X_\alpha, \mathcal{E}_\alpha) \in S$ and $f_\alpha : X \longrightarrow X_\alpha$. Then we consider the product $P_{\alpha<\lambda}(X_\alpha, \mathcal{E}_\alpha)$ and define $f : X \longrightarrow P_{\alpha<\lambda} X_\alpha$ by $f(x) = (f_\alpha(x))_{\alpha<\lambda}$. Since $\mathcal{K} \neq \mathfrak{M}_{single}$, f is injective and so we can identify X with $f(X)$ and consider the subspace $(X, \mathcal{E})$ of $P_{\alpha<\lambda}(X_\alpha, \mathcal{E}_\alpha)$. Clearly, $(X, \mathcal{E}) \in \mathbf{SP}\mathcal{K}$.

To see that $(X, \mathcal{E})$ is free with respect to $\mathcal{K}$, it suffices to verify that, for each $(X', \mathcal{E}') \in S$, every mapping $h : (X, \mathcal{E}) \longrightarrow (X', \mathcal{E}')$ is coarse. We take $\beta < \lambda$ such that $(X', \mathcal{E}') = (X_\beta, \mathcal{E}_\beta)$ and $h = f_\beta$. Then f_β is the restriction to X of the projection $pr_\beta : P_{\alpha<\lambda}(X_\alpha, \mathcal{E}_\alpha) \longrightarrow (X_\beta, \mathcal{E}_\beta)$. Hence, f_β is coarse. $\square$

Theorem 1. *For every class $\mathcal{K}$ of coarse spaces, the smallest variety, Var $\mathcal{K}$, containing $\mathcal{K}$ is* $\mathbf{QSP}\mathcal{K}$.

Proof. The inclusion $\mathbf{QSP}\mathcal{K} \subseteq \mathcal{K}$ is evident. To prove the inverse inclusion, we suppose that $\mathcal{K} \neq \mathfrak{M}_{single}$ (this case is evident) and take an arbitrary $(X', \mathcal{E}') \in Var(\mathcal{K})$. Then $(X', \mathcal{E}')$ can be obtained from $\mathcal{K}$ by means of some finite sequence of operations $\mathbf{S}, \mathbf{P}, \mathbf{Q}$. We use Lemma 2 to choose a coarse space $(X, \mathcal{E}) \in \mathbf{SP}\mathcal{K}$, with $|X| = |X'|$, which is free with respect to $\mathcal{K}$. By Lemma 1, any bijection $f : (X, \mathcal{E}) \longrightarrow (X', \mathcal{E}')$ is coarse so $(X', \mathcal{E}') \in \mathbf{QSP}\mathcal{K}$. $\square$

Theorem 2. *Let $\mathfrak{M}$ be a variety of coarse spaces such that $\mathfrak{M} \neq \mathfrak{M}_{single}$, and $\mathfrak{M} \neq \mathfrak{M}_{bound}$. Then there exists a cardinal κ such that $\mathfrak{M} = \mathfrak{M}_\kappa$.*

Proof. Since $\mathfrak{M} \neq \mathfrak{M}_{bound}$ and $\mathfrak{M} \neq \mathfrak{M}_{single}$, there exists a minimal cardinal κ such that $\mathfrak{M}$ contains an unbounded space of cardinality κ; so $\mathfrak{M} \subseteq \mathfrak{M}_\kappa$.

To verify the inclusion $\mathfrak{M}_\kappa \subseteq \mathfrak{M}$, we take a coarse space $(X, \mathcal{E}) \in \mathfrak{M}$, which is free with respect to $\mathfrak{M}$, and show that $(X, \mathcal{E})$ is free with respect to $\mathfrak{M}_\kappa$. We prove that $(X, \mathcal{E}) = (X, [X]^{<\kappa})$. If $|X| < \kappa$ then $(X, \mathcal{E})$ is bounded and the statement is evident. Assume that $|X| \geq \kappa$ but $(X, \mathcal{E}) \neq (X, [X]^{<\kappa})$. Assume that, for every ε, $\varepsilon = \varepsilon^{-1}$, the set $S_\varepsilon = \{x \in X : |B(x, \varepsilon)| > 1\}$ is bounded in $(X, \mathcal{E})$. By the choice of κ, $|S_\varepsilon| < \kappa$ and $|B(x, \varepsilon)| = 1$ for all $x \in X \setminus S_\varepsilon$. It follows that $(X, \mathcal{E}) = (X, [X]^{<\kappa})$. Then there exists an $\varepsilon \in \mathcal{E}$ such that the set S_ε is unbounded in $(X, \mathcal{E})$.

We choose a maximal, by inclusion, subset $Y \subset X$ such that $B(y, \varepsilon) \cap B(y', \varepsilon) = \varnothing$ for all distinct $y, y' \in Y$. We observe that Y is unbounded so $|Y| \geq \kappa$. We take an arbitrary $x_0 \in X$ and choose a mapping $f : X \longrightarrow X$ such that $f(y) = x_0$ for each $y \in Y$ and f is injective on $X \setminus Y$. Since $(X, \mathcal{E})$ is free with respect to $\mathfrak{M}$, the mapping $f : (X, \mathcal{E}) \longrightarrow (X, \mathcal{E})$ must be coarse. Hence, there exists an $\varepsilon' \in \mathcal{E}$ such that $f(B(x, \varepsilon)) \subseteq B(f(x), \varepsilon')$ for each $x \in X$. It follows that $f(\cup_{y \in Y} B(y, \varepsilon))$ is bounded in $(X, \mathcal{E})$. We note that $|f(\cup_{y \in Y} B(y, \varepsilon))| \geq \kappa$ so $(X, \mathcal{E})$ contains a bounded subset Z such that $|Z| = \kappa$. Since $(X, \mathcal{E})$ is free with respect to $\mathfrak{M}$, every $(X', \varepsilon') \in \mathfrak{M}$ is κ^+-bounded and we get a contradiction with the choice of κ. To conclude the proof, we take an arbitrary $(X, \mathcal{E}') \in \mathfrak{M}_\kappa$ and note that the identity mapping $id : (X, [X]^{<\kappa}) \longrightarrow (X, \mathcal{E}')$ is coarse so $(X, \mathcal{E}') \in \mathfrak{M}$. $\square$

Remark 1. *We note that $\mathfrak{M}_{single}$ is not closed under coarse equivalence because each bounded coarse space is coarsely equivalent to a singleton. Clearly, $\mathfrak{M}_{bound}$ is closed under coarse equivalence. We show that the same is true for every variety $\mathfrak{M}_\kappa$. Let $(X, \mathcal{E})$ be a coarse space, Y be a large subset of $(X, \mathcal{E})$. We assume that $(Y, \mathcal{E}|_Y) \in \mathfrak{M}_\kappa$ but $(X, \mathcal{E}) \notin \mathfrak{M}_\kappa$. Then X contains an unbounded subset Z such that $|Z| < \kappa$. We choose $\varepsilon \in \mathcal{E}$ such that $\varepsilon = \varepsilon^{-1}$ and $X = B(Y, \mathcal{E})$. For each $z \in Z$, we pick $y_z \in Y$ such that $z \in B(y_z, \mathcal{E})$. We let $Y' = \{y_z \in Z\}$. Since $|Y'| < \kappa$, Y' is bounded in $(Y, \mathcal{E}|_Y)$. It follows that Z is bounded in $(X, \mathcal{E})$, a contradiction with the choice of Z.*

We note also that every variety of coarse spaces is closed under formations of companions. For $\mathfrak{M}_{single}$ and $\mathfrak{M}_{bound}$, this is evident. Let $(X, \mathcal{E}) \in \mathfrak{M}_\kappa$ and $\mathcal{I}$ be the ideal of all bounded subsets of $(X, \mathcal{E})$. Since $(X, [X]^{<\kappa})$ is free with respect to $\mathfrak{M}_\kappa$, the identity mapping $id : (X, [X]^{<\kappa}) \longrightarrow (X, \mathcal{E})$ is coarse. Hence, $[X]^{<\kappa} \subseteq \mathcal{I}$ and $(X, \mathcal{E}) \in \mathfrak{M}_\kappa$.

Remark 2. *Every metric d on a set X defines a coarse structure $\mathcal{E}_d$ on X with the base $\{(x, y) : d(x, y) \leq n\}$, $n \in \omega$. A coarse structure $\mathcal{E}$ on X is called metrizable if there exists a metric d on X such that $\mathcal{E} = \mathcal{E}_d$. By ([3], Theorem 2.1.1), $\mathcal{E}$ is metrizable if and only if $\mathcal{E}$ has a countable base. From the coarse point of view, metric spaces are important in Asymptotic Topology, see [4].*

We assume that a variety $\mathfrak{M}$ of a coarse space contains an unbounded metric space (X, d) and show that $\mathfrak{M} = \mathfrak{M}_\omega$. We choose a countable unbounded subset Y of X and note that $(Y, d) \notin \mathfrak{M}_\kappa$ for $\kappa > \omega$ so $(Y, d) \in \mathfrak{M}_\omega \setminus \mathfrak{M}_\kappa$, and the variety generated by (X, d) is $\mathfrak{M}_\omega$.

3. Comments

1. Let G be a group with the identity e. An ideal $\mathcal{I}$ in $\mathcal{P}_G$ is called a *group ideal* if $[G]^{<\omega} \subseteq \mathcal{I}$ and $AB^{-1} \in \mathcal{I}$ for all $A, B \in \mathcal{I}$.

Let X be a G-space with the action $G \times X \longrightarrow X$, and $(g, x) \longmapsto gx$. We assume that G acts on X transitively, take a group ideal $\mathcal{I}$ on G, and consider the coarse structure $\mathcal{E}(G, \mathcal{I}, X)$ on X with the bases $\{\varepsilon_A : A \in \mathcal{I}, e \in A\}$, where $\varepsilon_A = \{(x, gx) : x \in X, g \in A\}$. Then $B(x, \varepsilon_A) = Ax$, where $Ax = \{gx : g \in A\}$.

By ([5], Theorem 1), for every coarse structure $\mathcal{E}$ on X, there exist a group G of permutations of X and a group ideal $\mathcal{I}$ in $\mathcal{P}_G$ such that $\mathcal{E} = \mathcal{E}(G, \mathcal{I}, X)$. Now let $X = G$ such that G acts on X by left shifts, $x \longmapsto gx$ for $g \in G$. We denote $(G, \mathcal{E}(G, \mathcal{I}, G))$ by $(G, \mathcal{I})$ and say that $(G, \mathcal{I})$ is a *right coarse group*. If $\mathcal{I} = [G]^{<\omega}$ then $(G, \mathcal{I})$ is called *a finitary right coarse group*. In the metric form, these structures on finitely generated groups play an important role in geometric group theory, see ([6], Chapter 4).

A group G endowed with a coarse structure $\mathcal{E}$ is a right coarse group if and only if, for every $\varepsilon \in \mathcal{E}$, there exists $\varepsilon' \in \mathcal{E}$ such that $(B(x, \varepsilon))g \subseteq B(xg, \varepsilon')$ for all $x, g \in G$. For group ideals and coarse structures on groups see ([3], Chapter 6) and [7].

2. A class $\mathfrak{M}$ of right coarse groups is called a *variety* if $\mathfrak{M}$ is closed under formation of subgroups, coarse homomorphic images, and products.

Let $\mathcal{K}$ be a class of right coarse groups, and G be a group generated by a subset $X \subset G$. We say that a right coarse group $(G, \mathcal{I})$ is *free* with respect to $\mathcal{K}$ if, for every $(G', \mathcal{I}') \in \mathcal{K}$, any mapping $X \longrightarrow G'$ extends to the coarse homomorphism $(G, \mathcal{I}) \longrightarrow (G', \mathcal{I}')$. Then Lemmas 1 and 2 and Theorem 1 hold for the right coarse groups in place of coarse spaces.

Let $\mathfrak{M}$ be a variety of right coarse groups. We take an arbitrary $(G, \mathcal{I}) \in \mathfrak{M}$, delete the coarse structure on G and the class $\mathfrak{M}^\flat$ of the groups. If $(G, \mathcal{I}) \in \mathfrak{M}$ then $(G, \mathcal{P}_G) \in \mathfrak{M}$. It follows that $\mathfrak{M}^\flat$ is a variety of groups.

Now let $\mathcal{G}$ be a variety of groups different from the variety of singletons. We denote by $\mathcal{G}_{bound}$ the variety of right coarse groups $(G, \mathcal{P}_G)$, *for* $G \in \mathcal{G}$. For an infinite cardinal κ, we denote by $\mathcal{G}_\kappa$, the variety of all κ-bounded right coarse groups $(G, \mathcal{I})$, for $G \in \mathcal{G}$.

Let $\mathfrak{M}$ be a variety of right coarse groups such that $\mathfrak{M}^\flat = \mathcal{G}$. In contrast to Theorem 2, we do not know if $\mathfrak{M}$ lies in the chain:

$$\mathcal{G}_{bound} \subset \ldots \subset \mathcal{G}_\kappa \subset \ldots \subset \mathcal{G}_\omega.$$

If G is a group of cardinality κ and $G \in \mathcal{G}$ then $(G, [G]^{<\kappa}) \in \mathcal{G}_\kappa \setminus \mathcal{G}_{\kappa'}$ for each $\kappa' > \kappa$. Hence, all inclusions in the above chain are strict.

3. Let Ω be a signature, A be an Ω-algebra, and $\mathcal{E}$ be a coarse structure on A. We say that A is a *coarse Ω-algebra* if every n-ary operation from Ω is coarse, for example the mapping $(A, \mathcal{E})^n \longrightarrow (A, \mathcal{E})$. We note that each coarse group is a right coarse group but the converse statement need not be true, see ([3], Section 6.1).

A class $\mathfrak{M}$ of a coarse Ω-algebra is called a *variety* if $\mathfrak{M}$ is closed under formation of subalgebras, coarse homomorphic images, and products. Given a variety $\mathfrak{M}$ of coarse algebras, the class $\mathfrak{M}^\flat$ of all Ω-algebras A, such that $(A, \mathcal{E}) \in \mathfrak{M}$, is a variety of Ω-algebras. Let $\mathcal{A}$ be a variety of Ω-algebras different from the variety of singletons. We let $\mathcal{A}_{bound}$ be the variety of coarse algebras $(A, \mathcal{P}_A)$, for $A \in \mathcal{A}$. For an infinite cardinal κ, we denote $\mathcal{A}_\kappa$ as the variety of all κ-bounded Ω-algebras $(A, \mathcal{E})$ such that $A \in \mathcal{A}$, and get the chain:

$$\mathcal{A}_{bound} \subseteq \ldots \subseteq \mathcal{A}_\kappa \subseteq \ldots \subseteq \mathcal{A}_\omega,$$

however, we can not state that all inclusions are strict. In the case of course groups, this is because each non-trivial variety of groups contains some Abelian group A, of cardinality κ, and the coarse group $(A, [A]^{<\kappa})$ is κ-bounded but not κ^+-bounded.

4. A class $\mathfrak{M}$ of topological Ω-algebras (with regular topologies) is called a *variety (a wide variety)* if $\mathfrak{M}$ is closed under formation of closed subalgebras (arbitrary subalgebras), continuous homomorphic images, and products. Wide varieties and varieties are characterized syntactically by the limit laws [8] and filters [9]. In our coarse case, the part of filters is played by the ideals $[X]^{<\kappa}$.

There are only two wide varieties of topological spaces, the variety of singletons and the variety of all topological spaces, but there are plenty of varieties of topological spaces. The variety of coarse spaces $\mathfrak{M}_\kappa$ is a twin of the varieties of topological spaces in which every subset of cardinality $< \kappa$ is compact. We note also that $\mathcal{G}_\kappa$ might be considered as a counterpart to the variety $T(\kappa)$ of topological groups from [10], where $G \in T(\kappa)$ if and only if each neighborhood of e contains a normal subgroup of index strictly less then κ.

5. A class $\mathfrak{M}$ of uniform spaces is called a *variety* if $\mathfrak{M}$ is closed under formation of subspaces, products and uniformly continuous images. For an infinite cardinal κ, a uniform space X is called κ-bounded if X can be covered by $< \kappa$ balls of arbitrary small radius. Every variety of uniform spaces different from varieties of singletons and all spaces coincides with the variety of κ-bounded spaces for some κ, see [11]. I thank Miroslav Hušek for this reference.

Conflicts of Interest: The authors declare no conflict of interest.

References

1. Roe, J. *Lectures on Coarse Geometry*; University Lecture Series; American Mathematical Society: Providence, RI, USA, 2003; Volume 31, p. 176.
2. Protasov, I.; Banakh, T. *Ball Structures and Colorings of Groups and Graphs*; Mathematical Studies Monograph Series; VNTL Publishers: Lviv, Ukraine, 2003; Volume 11.
3. Protasov, I.; Zarichnyi, M. *General Asymptology*; Mathematical Studies Monograph Series; VNTL Publishers: Lviv, Ukraine, 2007; Volume 12, p. 219.
4. Dranishnikov, A. Asymptotic Topology. *Russ. Math. Surv.* **2000**, *55*, 1085–1129. [CrossRef]
5. Petrenko, O.V.; Protasov, I.V. Balleans and G-spaces. *Ukr. Math. J.* **2012**, *64*, 387–393. [CrossRef]
6. Harpe, P. *Topics in Geometrical Group Theory*; University Chicago Press: Chicago, IL, USA, 2000.
7. Protasov, I.V.; Protasova, O.I. Sketch of group balleans. *Math. Stud.* **2004**, *22*, 10–20.
8. Taylor, W. Varieties of topological algebras. *J. Aust. Math. Soc.* **1977**, *23*, 207–241. [CrossRef]
9. Protasov, I. Varieties of topological algebras. *Sib. Math. J.* **1984**, *25*, 783–790. [CrossRef]
10. Morris, S.; Nickolas, P.; Pestov, V. Limit laws for wide varieties of topological groups II. *Houst. J. Math.* **2000**, *26*, 17–27.
11. Pelant, J.; Ptak, P. On σ-discreteness of uniform spaces. In *Seminar Uniform Spaces 1975–1976*; Frolik, Z., Ed.; Mathematical Institute of the Czechoslovak Academy of Sciences in Prague: Praha, Czech Republic, 1978; pp. 115–120.

MDPI

St. Alban-Anlage 66

4052 Basel

Switzerland

Tel. +41 61 683 77 34

Fax +41 61 302 89 18

www.mdpi.com

Axioms Editorial Office

E-mail: axioms@mdpi.com

www.mdpi.com/journal/axioms

www.ingramcontent.com/pod-product-compliance
Lightning Source LLC
Chambersburg PA
CBHW051904210326
41597CB00033B/6023